Your Hospitality
Field Experience

Your Hospitality Field Experience

A Student Workbook

Jack E. Miller, F.M.P.

and Karen Eich Drummond, F.M.P., R.D.

John Wiley & Sons, Inc.

New York • Chichester • Brisbane • Toronto • Singapore

Library of Congress Cataloging in Publication Data:

Miller, Jack E., 1930–
    Your hospitality field experience :   a student workbook / by Jack E. Miller and Karen Eich Drummond.
      p.   cm.
    Includes bibliographical references.
    ISBN 0-471-05327-9 (pkb.: acid-free paper)
    1. Hospitality industry—Study and teaching. 2. Occupational training.
 I. Drummond, Karen Eich. II. Title.
    TX911.5.M55 1995
    647'.0715—dc20   95-12109

Printed in the United States of America

10 9 8 7 6 5 4 3 2 1

Dedicated to you, the hospitality student. We hope that by using this book you will learn more about the hospitality field in general and where you might enjoy working in particular.

Preface

Y*our Hospitality Field Experience* is designed for use by college students who will be completing a *field experience* (a term that encompasses internships, co-ops, practicums, and externships) in hotels, restaurants, or institutions.

In Chapter 1, students are acquainted with the procedures involved in finding a hospitality job, which may provide the setting for the field experience. Many how-tos are covered, including defining career goals, locating prospective employers, writing a resume and cover letter, taking interviews, and doing an outstanding job for the employer. Career goal worksheets are presented along with samples of resumes, cover letters, and thank-you letters.

The remaining chapters contain Training Goals, Training Guides, and Introductory Worksheets, as follows.

Chapter 2. Training Goals and Training Guides for Hotels

Chapter 3. Training Goals and Training Guides for Restaurants and Foodservices

Chapter 4. Introductory Worksheets for Institutional Foodservices

Chapter 5. Training Goals and Training Guides for Hospitality Human Resource Management

Instead of being categorized by job titles, **Training Goals** are given for different functions, such as table service or purchasing. The premise is that irrespective of the size or geographic location of the operation, persons accomplish certain common job functions or tasks. The procedures, methods, or competency levels may vary greatly from organization to organization, but the commonality will continue to exist.

For example, it can be assumed that every desk clerk will register a guest; there are three components to this registration: greeting, completion of the registration record, and distribution of registration information. The student who can accomplish and understand the need for these three components of registration regardless of the corporate procedural requirements or methodology will be equipped to adjust more quickly to changes and/or differences in procedures in his own or a different company.

The purpose of the Training Goals is to ensure that every major job function is included in the student's work experience.

Training Guides list job-related questions for the intern to answer either on or off the job. The purpose of the Training Guides is to

1. Help focus the student's learning.
2. Involve job supervisors in a meaningful instruction process by serving as a reference source of requested information.
3. Help students expand their knowledge of the departments where they may be able to spend only a little, or no, time.
4. Help students correlate theory and operational procedures in a work situation.
5. Stimulate student interest and participation.
6. Serve as springboards for class discussion, especially as a way to compare different hospitality operations and their procedures and operating methods.
7. Serve as a means to help evaluate student performance.

The Training Guides, when used as an individual instruction tool, facilitate a specialized form of learning activity. It is for the total scheme of the student's education that these guides have been developed. They contribute to the classroom, theoretical, and practical areas of education. These guides are a learning activity for the student and will coordinate the school with the work situation.

Introductory Worksheets are designed for students to use when they first start their field experience. The following worksheets are available in this book.

Introduction to Hotels (beginning of Chapter 2)

Introduction to Restaurants and Foodservice (beginning of Chapter 3)

Introduction to Business and Industry Foodservices (Chapter 4)

Introduction to School Foodservice (Chapter 4)

Introduction to Healthcare Foodservice (Chapter 4)

Introduction to College and University Foodservice (Chapter 4)

Introduction to Retirement Community Foodservice (Chapter 4)

Introduction to Human Resource Management (beginning of Chapter 5)

The purpose of these worksheets is for students to first see the "big picture" of the operation in which they are working. For example, in "Introduction to Restaurants and Foodservices," students are asked questions about the mission of the restaurant or foodservice, the organizational structure, the menu, the facility's layout and design, and other areas.

If a student is working in one of the five institutional settings discussed in Chapter 4, he or she should first fill out the "Introduction to Foodservice" worksheet in Chapter 2, then complete the worksheet for the specific setting found in Chapter 4. The worksheets focusing on the institutional settings delve into areas specific to those environments.

The benefits of this book include the following:

1. The first chapter gives practical information to help students find a field experience employer. This is a topic that cannot always be covered in class.
2. The Training Goals make it clear to students what they need to be learning on the job.
3. The Training Guides help focus the student's learning, help the student correlate theory and operational procedures in a work situation, involve job supervisors, stimulate student interest and participation, serve as springboards for class discussion, and help in student evaluation.

What we feel is particularly beneficial in this book is its depth, its Introductory Worksheets, and the fact that the Training Goals and Training Guides can be customized by the instructor. Extra spaces are given at the bottom of each Training Goals and Training Guide sheet in which instructors can ask students to insert any additional goals or questions to answer.

We trust this workbook will be as helpful as we have intended. To both students and instructors, we wish good luck!

Jack E. Miller
Karen Eich Drummond

Acknowledgments

The authors wish to thank Dr. John E. Elias, Coordinator, Distributive Teacher Education, College of Education, University of Missouri, Columbia, Missouri, who challenged Jack E. Miller to solve the problems this book addresses. Thanks also to the many students of St. Louis Community College at Forest Park enrolled in the Hotel, Restaurant and Institutional Program who participated in discussions of their work experiences and employment needs. Karen Eich Drummond would like to thank her co-author for suggesting that she be involved in revising this book. She would also like to thank her many past and present students, who make teaching a two-way street.

Contents

Your Hospitality
Field Experience

Getting a Job

M ost hospitality and tourism education programs involve some field experience or working a job in the field. Field experiences come by many names (such as internship, externship, co-op, or practicum) and are probably administered somewhat differently by each school offering them. In any case, field experiences do have several common elements: You work a job in which you apply theory to real work situations while under the supervision of an on-site hospitality manager as well as a faculty member.

The purpose of this chapter is to acquaint you, the student, with the procedures involved in finding a job that may be part of your school's field experience. Securing employment is a skill you will need throughout your life. This chapter will also give you some tips once you start working. A great deal of your education is going to take place outside of the classroom. This part of your education will be extremely important as a future full-time employee in the hospitality industry, in positions that will be based on the experiences and skills that you acquire from work situations. Your on-the-job training—the responsibilities that you take on and the business problems that you become familiar with—will be as important a part of your education as what is learned in the classroom.

Setting Career Goals

Before looking for a job, you need to take some time to assess yourself as a present and future employee. What type of job is best suited to someone with your temperament, talents, and skills? What type of work do you want to do right now? In the future? What job-related experience do you want to gain? What you will be doing is an analysis of yourself, and it is a very important part of your job search. Nobody can answer these questions about yourself but you, so what follows is a procedure that you may want to try.

Set up a sheet of paper with the five headings shown in Figure 1.1. First, make a list of what you consider to be your strongest skills. For example, you may feel that you are very good at waiting on tables or at balancing out a cash register at the end of the day. Next, list the different types of work experience that you have gained from the jobs you previously held or currently hold. These two lists combined should give you a fairly accurate picture of the skills that you possess that make you an "employable" product. At the same time, consider the availability of hospitality employment in your geographic area for someone of your age and with your abilities.

Now, list any skills in which you are not overly competent but that you would

---

Strongest Skills

1. Willingness to learn
2. Physical stamina
3. Availability at all hours
4. Dining room experience
5. H&R education

Work Experience

1. Dishwasher
2. Busboy
3. Prep cook

Needed Competence

1. Kitchen experience under actual conditions in a full-service restaurant
2. Production planning and management
3. Skill training by competent chefs
4. Broiler work
5. Fry cook work
6. Soup and sauce preparation

Likes and Dislikes

LIKE Food preparation
LIKE Gourmet cooking
DISLIKE Working in dining room
DISLIKE Daily morning hours

<u>Short-Term Objective:</u> To obtain full-time summer employment in a full-service restaurant with high food standards; to work various stations under the direction of the chef and other experienced employees.

Figure 1.1 Lists help you develop short-term and long-term career goals

like to develop. Also, list any experiences that you have not had in previous work situations that you would like to have now or that you feel may be useful to you in the future. In the same way, make a list of what you like and do not like to do, for your interests are certainly as important a consideration as anything else when you are after the right job.

These four lists—what you are good at, what you are experienced at, what you want to learn, and what you like to do—will help you to define more clearly the type of job to which you are best suited. They will help you to begin developing short-term and long-term career goals for yourself. Of course, it is unrealistic to assume that as a student you will be able to make absolute decisions in regard to long-range career goals. But if you can think of yourself in terms of what you like to do, what you want to do, and what you are good at, and shape your plans for first employment accordingly, then you have in fact begun to assess your own goals and given yourself a direction in which to head.

Roughly a long-term goal is the place that you eventually want to be at in the hospitality field. A short-term goal is the job that you would like to have in the immediate future as well as the job that will properly prepare you for where you want to be in the years to come. Both goals essentially involve determining your needs from employment. Consider this example: Suppose that you have never tended bar before but think that you would like to have a job as a bartender. Why do you feel that tending a bar would be an appropriate and enjoyable job for you? Would the hours, salary, and job duties of a bartender satisfy your immediate needs for employment? If you can answer these questions to your satisfaction, then you have singled out a short-term goal.

Now consider this question: What kinds of learning experiences will a bartender's job offer you that may be useful to you in the future? Suppose you feel that within ten years you may want to own your own bar. What do you need to know in order to achieve this goal? Obviously, you are going to have to know how to open up a bar and how to operate it. But you will also have to know how to purchase liquors as well as something about personnel supervision. Being a bartender is one way in which to prepare yourself for the experience of owning your own bar. But what other types of employment will prepare you for this experience in ways in which a bartender's job cannot? Have you acquainted yourself with the various companies in your area that will be best able to equip you with the experience that you need? Knowing what you need to know in order to get to where you eventually want to be is the type of thinking that characterizes long-term goal planning.

As you begin to think about both your short- and long-term goals, you will always want to keep the nature of those goals in mind. That is, you want to be sure that the career goals that you have selected for yourself are realistic. If you have analyzed yourself successfully, then your general abilities should be reasonably clear to you. Set up your goals using what you know about yourself and your abilities as a guideline, and your chances for success will at least be considerably increased. People who set up goals for themselves that they cannot possibly attain are in fact setting themselves up to fail. They have mapped out a plan that, for one reason or another, is impossible for them to carry out. As a result, they will never really feel successful because they will never accomplish all of the tasks that they have decided upon. Be honest with yourself, and be shrewd enough to choose and plan a career that will make extensive use of your strongest points.

If you are having trouble deciding on a short-term career goal, you might do well to set up an appointment with your faculty advisor or another faculty member. Your instructors are thoroughly familiar with the field that you are entering and should be able to give you a clear idea of what types of skills your chosen career goal involves. They may suggest to you a number of occupations for which they feel you are well suited. Or their comments and explanations may steer you away from a job that you thought you wanted. In any event, whatever advice they have to offer can only be of benefit to you.

Locating Prospective Employers

After you have examined your employable assets as well as your weaknesses and decided upon what type of job you are going to look for, you must begin to think about locating prospective employers. Your school's placement service may be able to help you to learn more about the companies that interest you. Most schools maintain a fairly detailed file on the companies in their general geographic region and on those that come to the school to recruit. The information in these files may be in the form of various brochures and fliers that the company sends out in hopes of attracting potential employees.

Another good source of employment will probably be through the hospitality and foodservice departments of your school, for they are likely to receive numerous calls from employers in the area who are looking for students to fill any part-time or full-time openings. Some departments maintain a bulletin board with job postings.

Perhaps the easiest way to find a job is simply to ask fellow students if they know of any openings around town or where they themselves are employed. Not only can your classmates tell you about unfilled jobs at their own companies (which you may not be aware of otherwise), they can also give you a fairly accurate idea of what type of organization their company is and what type of people work there.

Another resource is hospitality conventions and trade shows, such as the American Hotel & Motel Association convention held in New York City every November, the National Restaurant Association Convention held in Chicago every May, and local trade shows. These shows offer wonderful opportunities to learn about and meet with hospitality companies that may be hiring.

In addition to talking to people, written advertisements are a source of information. Newspaper classified and display advertisements can supply local job leads; Sunday newspapers have the most advertising. For at least two weeks, read the classified and display advertisements from A to Z to get an idea of what types of hospitality jobs are available. Many jobs are advertised under "Hotel" or "Restaurant," but you will see other hospitality jobs in between.

Trade magazines, such as *The Nation's Restaurant News*, always contain job advertising. Be aware, however, that the jobs listed in them are going to be primarily full-time and may not be located in your area. Another good magazine to look at is the student magazine, *Hosteur* ™, which is written for students in colleges and universities offering hospitality and tourism programs.

No discussion about getting a job would be complete without discussing "networking." Networking is the informal sharing of information among individuals or groups with a common interest, such as advancing one's career. Networking with hospitality representatives can provide you with job and career information and advice, and possibly a job interview or referral to another company that has openings.

Whenever you have the opportunity to meet with hospitality representatives, such as at a career fair or hospitality reception, take the time to introduce yourself to them and ask them about themselves, their work, and their company. Don't just start telling the person right away how you are looking for a job, how great an employee you are, or that you would like his business card for future reference. Wait until you have built up a rapport with the individual before you ask if you can contact him in the future to discuss career goals, at which point, hopefully, the person will offer you a business card.

Now, it may seem that going to a funeral would be more fun than approaching people at a packed hospitality reception. It is true that networking at such a reception is a difficult skill to learn, but it can be learned! First, forget that old-fashioned notion that you have to have someone introduce you to somebody else. It's perfectly acceptable to walk up to a total stranger, extend your hand, smile, and introduce yourself. Get the conversation going by asking the person about her company and her role in that company. Keep a sense of humor and avoid topics such as politics, college pranks, religion, and how much you dislike your current job.

How long you talk with someone will depend on your judgment. One school of thought says to meet as many people as you can during a reception and collect their business cards. Another school of thought says to strike up just two or three quality conversations. In any case, once you sense that the conversation needs to be ended, tell the person, "It was a pleasure meeting you, Mr. So-and-So," and move on. After this type of networking opportunity, send a cover letter and resume to those people who you think can help in your job search.

Writing A Resume

After you have started looking for places of employment, you also need to think about developing a resume. The resume is a brief, factual statement of who you are, what you can do, and what experiences you have had. When you write up a resume, keep your prospective employer in mind; you will be informing the employer about yourself as an individual. Do not underestimate the importance of your resume. It is one of the most critical parts of securing a job. Remember that your resume precedes the interview and that it is the only impression you make on your potential employer before, hopefully, you are asked to make a personal impression in an interview.

It is essential, therefore, that you present yourself to your best advantage on your resume. In all cases the resume should be professionally done. That is, it should be printed for each company to which it is sent. Resumes should be typeset or, if printed by computer, printed on a laser printer. Those that are sloppily done create a bad impression instantly and can prevent you from obtaining an interview with the company of your choice. Think of your resume as your advertising vehicle for yourself. In seeking or in selecting employment, the only product that you have to sell is YOU. Therefore, you will want to make your resume the best sales vehicle that you possibly can. If it is a well-written statement, printed professionally on high-quality paper, then it is going to make that much better an impression.

Try to be brief on the resume. Your prospective employer must be given enough detail about you to be able to assess the extent of your capabilities on the basis of your education and your work experience. However, do not "pad" your resume with unnecessary information. For instance, the recruiter needs to know about the different jobs that you have held in the past few years. He or she also needs to know about the different duties that you have performed in those jobs.

But your recruiter needs only a minimum of information about those jobs that have no relation to the position for which you are applying. Your resume will be easier for the recruiter to read and absorb if it is a compact document.

Do not leave spaces of time on your resume. A company will notice any gaps on your resume and if they do choose to interview you, they will certainly ask you what you were doing during that time. Even if you were working in another field altogether, it's okay to put that on your resume. In some cases, it even works to your advantage.

Be accurate and honest on your resume. It is the tendency of most people to exaggerate on certain parts of their education or experience. Naturally you will want to present yourself in the best possible light, and it is reasonable of you to want to play up your strengths. But it is possible to have a positive one and an honest resume at the same time.

Now for the nuts and bolts of resume writing. Although there are many schools of thought on what to include, experts agree that work experience and education are important on any resume. If you have any relevant professional affiliations, awards and honors, or special skills, they should be stated as well. Another area you might include, as space permits, is community service experience.

Figure 1.2 shows a sample resume worksheet. On this sheet you will gather the necessary information from which to write up your final draft. Don't skip this step—it makes it easier and less frustrating to write up your resume because you will have everything at your fingertips.

Now you are ready to write up your resume. First, you need to center your name, full address, and telephone number at the top of the page. If you don't already have an answering machine on your telephone, you may want to think about making this wise investment.

Some people start off their resume with an objective, basically a short statement of what type of job you are looking for. It is important that the objective be concise and not too broad or specific. For instance, if you are looking for a dining room job, you might state: "Dining room position serving guests." Some applicants like to use an objective; others don't. It's a personal decision. The information stated in your objective is going to be stated in your cover letter, so it is not absolutely essential that it be on the resume. If you do decide to keep the objective on your resume, check that what you have stated is appropriate each time you send your resume out.

The first section of your resume will contain either your work experience or your education. A good rule of thumb is to put the one that will be seen as stronger by the recruiter. If you start with work experience, always put your most recent job first and work your way backward. As you can see in Figure 1.3, you should include your job title, your place of employment and the town in which it is located, and your dates of employment. Mentioning both month and year is a good idea when you have only worked at a job for a year or two or less.

You can summarize what you did at each job either by writing a short paragraph summarizing your major job duties and accomplishments or by listing them as seen in Figure 1.3. You will note that each job duty or accomplishment listed in the Figure starts with a specific, descriptive verb. Instead of using general verbs such as "manage" or "oversee," use precise verbs such as "organize" or "budget." Figure 1.4 lists action verbs that you can use when preparing your resume.

In the next section you will discuss your education. It is important to include here the names of colleges and universities, dates of attendance or date of graduation, and the type of degree received or your major if you are still working on your degree. If your cumulative average is particularly good or you had any honor or

OBJECTIVE

Practice writing a job objective until you've written one that best reflects your short-term career goals.

_____

_____

_____

_____

_____

_____

_____

_____

_____

EDUCATION

Use the following worksheet to write down information about the schools you have attended. Start with the most recent school and work backwards.

School: _____

Major: _____

Date Graduated/Dates Attended: _____   Degree Seeking or Granted: _____

Grade Point Average: _____

Offices Held: _____

Scholarships: _____

_____

Honors: _____

_____

Extracurricular Activities: _____

_____

_____

___

Figure 1.2   Resume Worksheet

Continues

School: _____

Major: _____

Date Graduated/Dates Attended: _____   Degree Seeking or Granted: _____

Grade Point Average: _____

Offices Held: _____

Scholarships: _____

_____

Honors: _____

_____

Extracurricular Activities: _____

_____

School: _____

Major: _____

Date Graduated/Dates Attended: _____   Degree Seeking or Granted: _____

Grade Point Average: _____

Offices Held: _____

Scholarships: _____

_____

Honoraries: _____

_____

Extracurricular Activities: _____

_____

School: _____

Major: _____

Date Graduated/Dates Attended: _____   Degree Seeking or Granted: _____

Grade Point Average: _____

Offices Held: _____

Scholarships: _____

_____

Honors: _____

_____

Extracurricular Activities: _____

_____

Figure 1.2 Resume Worksheet *(Continued)*   *Continues*

WORK EXPERIENCE

Use the following worksheet to write down information about the jobs you have had. Start with your current job or most recent job, then work backward.

Dates Employed: _____
(use month/year preferably)

Employer's Name: _____

Employer's Address: _____

_____

Job Title: _____

Major Job Duties: _____

_____

_____

_____

_____

Accomplishments: _____

_____

_____

Dates Employed: _____
(use month/year preferably)

Employer's Name: _____

Employer's Address: _____

_____

Job Title: _____

Major Job Duties: _____

_____

_____

_____

_____

Accomplishments: _____

_____

_____

Figure 1.2 Resume Worksheet *(Continued)*   *Continues*

Dates Employed: _____
(use month/year preferably)

Employer's Name: _____

Employer's Address: _____

_____

Job Title: _____

Major Job Duties: _____

_____

_____

_____

_____

Accomplishments: _____

_____

_____

Dates Employed: _____
(use month/year preferably)

Employer's Name: _____

Employer's Address: _____

_____

Job Title: _____

Major Job Duties: _____

_____

_____

_____

_____

Accomplishments: _____

_____

_____

Figure 1.2 Resume Worksheet *(Continued)*

Continues

Dates Employed: _____
(use month/year preferably)

Employer's Name: _____

Employer's Address: _____

_____

Job Title: _____

Major Job Duties: _____

_____

_____

Accomplishments: _____

_____

REFERENCES

Be sure to get the following information for three references.

1. Name: _____

   Address: _____

   _____

Telephone: _____

Job Title (if employed): _____

Type of Reference: _____

2. Name: _____

   Address: _____

   _____

Telephone: _____

Job Title (if employed): _____

Type of Reference: _____

3. Name: _____

   Address: _____

   _____

Telephone: _____

Job Title (if employed): _____

Type of Reference: _____

Figure 1.2   Resume Worksheet *(Continued)*

Continues

Donna Orange
999 Longwood Road
Surfs Up, CA 12345
(415) 415-4151

OBJECTIVE: Position serving guests in hotel dining room.

WORK EXPERIENCE

7/92–Present   **Dining Room Server**, Pizza Palace (Surfs Up, CA)

- Perform opening duties.
- Greet and seat customers.
- Assist customers in food selection.
- Take customers' orders.
- Serve food and beverages.
- Handle payment.
- Perform sidework and closing duties.
- Received "Outstanding" performance evaluation on last review.

6/90–7/92   **Counter Server**, Burger Bear (Surfs Up, CA)

- Took and filled customer orders.
- Took payment.
- Cleaned counter and dining areas.

EDUCATION

9/92–Present   State University, San Francisco, CA
Junior in Hotel and Restaurant Program
Grade Point Average: 3.5
Membership Chair, Society of Hosteurs

12/94   SERVSAFE Serving Safe Food Certificate

REFERENCES AVAILABLE UPON REQUEST.

Figure 1.3   Sample Resume

Clerical skills	formulated	coordinated
arranged	persuaded	developed
catalogued		directed
compiled	**Financial skills**	evaluated
generated	administered	improved
organized	analyzed	supervised
processed	balanced	
systematized	budgeted	**Research skills**
	forecast	clarified
Creative skills	marketed	evaluated
conceptualized	planned	identified
created	projected	inspected
designed		organized
established	**Helping skills**	summarized
fashioned	assessed	
illustrated	coached	**Technical skills**
invented	counseled	assembled
performed	diagnosed	built
	facilitated	calculated
Communication skills	represented	designed
arranged		operated
addressed	**Management skills**	overhauled
authored	administered	remodeled
drafted	analyzed	repaired

Figure 1.4   Action Verbs for a Resume

scholarships, you can mention them, too. If you were involved in any hospitality-related clubs or associations, especially if you held an office, these are good to put down on a resume as well. To round out your education section, you should mention any relevant certificates you have, such as a Certified Working Chef or Certified Dietary Manager, as well as any substantive training courses that you may have completed with past or present employers.

If you have been in the military, it is appropriate at this point to put in a section on your military background. You should include at least your branch of service, when you served, and in what capacity you served.

At this point in a resume, some applicants put in personal data, such as height and marital status. Such information has no relevance to any jobs. Therefore, your resume should not include your age or birthdate, marital status, health status, or other personal information that is not job-related.

The final section of your resume should state something like this: "References available upon request." A reference can be defined as someone whom your prospective employer may feel free to contact in order to find out more about you. He would contact a personal reference to find out more about your character and a professional reference to find out more about your work habits and your abilities. In most cases, your interviewer will expect you to present him with at least three references. Try to include an educator as one reference and a friend of the family or member of the clergy as another. Also include a former employer.

There are a few important points to keep in mind while gathering your references. Only use references who are going to give a positive report about you. Moreover, do not use their names on application forms or on the phone without first obtaining their permission. Either by phoning or by writing a letter, explain to them that you are looking for a job and would like to be able to use their name as a reference. Supply your references with ample information about yourself and what kind of job you are looking for. You will be helping yourself by giving your references some idea of the information that your potential employer(s) will want. Finally, courtesy is important; once you have been placed in a job, write to the people who served as your references, letting them know that you have found a job and thanking them for their help.

Now that you have a resume in hand, you may be wondering if it's too long. For students without an extensive work history, your resume will most probably fit onto one page. If your experience is extensive, or particularly impressive, you can certainly have a two-page resume, but try to keep it at that. Remember that the person reading it is probably going to zero in on the highlights of your career, so keep your resume concise.

Following is a resume preparation checklist. Use it to fine-tune your resume.

1. Be concise, keep your resume to one page. A two-page resume is okay if you have extensive experience, such as over 10 years.
2. Be accurate and honest. Don't ever think that recruiters won't check facts on your resume because they will, using one method or another.
3. Have your resume professionally printed. Use white paper or a very pale color if you don't want white. Ivory or pale gray is generally acceptable. Don't pick hot pink paper in the hopes of attracting attention. You will no doubt attract attention, but you will probably also get a quick trip to the "reject" bin.
4. Don't leave gaps in your work experience.
5. Don't include personal information.
6. Do get permission from references ahead of time and have their addresses and phone numbers available.
7. Include your salary history only if the employer you are sending your resume to requests this information.
8. Proofread your resume carefully. Ask a friend to proofread it too.

Once your resume is completed, keep it up to date. It is a good idea to revise it every two years or so, even if you are not planning a job change. Your resume is meant to be an account of the professional part of your life, and as you continue to work, your profession and your place in it continue to change. Your duties may change, your location may change, the people who serve as your references may change addresses, or you may not be able to or may not want to keep them as references. It is especially important to keep the names and addresses of your references up to date.

One last note about resumes. Even though you have a meticulous resume, you will often be asked by employers to fill out an employment application. It is simply many employers' policy to require applications of all applicants, even those submitting resumes, so don't fight it. Just fill out the application form using these guidelines:

1. Use a black or blue pen.
2. Answer every question. If a question does not apply to you, write "Not applicable" in the blank space.
3. Be neat. Printing is probably preferable.
4. Don't abbreviate.
5. Sign your full name and date the form.

Writing a Cover Letter

The cover letter is a brief business letter addressed to the company to which you are applying. The letter should capture the employer's attention, show why you are writing, indicate why your employment will benefit the company, and ask for an interview. The letter accompanies the resume and can be written either in response to a specific job opening or as a letter inquiring about the possibility of an opening. It is above all brief and positive.

Like the resume, the cover letter is essentially a direct-mail sales vehicle. Because both the resume and the cover letter precede the interview, you are relying on the written word to introduce you to the company. Whether this company decides to interview you will depend upon the extent to which your resume and cover letter appealed to them.

Your cover letter is meant to leave your prospective employers with the impression of what you have to offer them rather than of what you want to take from them. What skills do you have that make you a suitable candidate for the job? Why would you make a valuable employee for this company? This is the information that you want to communicate to the employer. You do not want to spend much time telling your employer that the job would be a "good learning experience" for you. Of course, you want a job that will provide you with the opportunity to learn some of the skills that you lack, but keep in mind that the company is hiring you for what you can do for them.

The typical cover letter (Figure 1.5) has three paragraphs. The first paragraph should state the position you are applying for and how you heard about the job opening, such as through an advertisement or from someone who works there. It is also important to make a pitch here about what you know about the employer and what you can do for them. By reading the business section of local newspapers or industry publications, you may find that a company recently received an award or favorable comments. The second paragraph should briefly stress your qualifications for the job and make reference to your resume. In the third paragraph you should request an interview, give contact information, and thank the recruiter for his or her consideration.

Tips to keep in mind here include the following:

1. Maintain a professional demeanor throughout your letter.
2. Make your letter positive and upbeat.
3. Your letter should have a professional appearance, and the typeface should match that of your resume.
4. Maintain a self-confident attitude throughout your letter; remember, you are applying for a job with the intention of getting it.
5. Be sure to address your letter to the appropriate person in the company. You can find this out by calling the human resources or personnel department.

999 Longwood Road
Surfs Up, CA 12345
January 5, 1995

Mr. John White
Director, Food and Beverage
Stately Hotel
111 Stately Avenue
San Francisco, CA 12346

Dear Mr. White,

Don Black suggested that I send my resume to you in order to be considered for the part-time Dining Room Server position that is open. I understand your dining room to be a first-class operation and believe that my skills and background will fit your needs.

During the past two years I have worked as a dining room server at the Pizza Palace. In addition to working, I attend classes in the Hotel and Restaurant program at the State University in San Francisco. I am presently a junior and have completed coursework in dining room service.

I am excited about the opportunity of discussing with you the possibility of my serving in your restaurant. I am available most weekdays for an interview, and you can call me at the phone number below. Thank you for your consideration.

Sincerely,

Donna Orange
(415) 415-4151

Figure 1.5   Sample Cover Letter

Now you have in your hands a great little resume. What do you do? Well, there are basically two approaches:

1. Send the same cover letter and resume to any and every hospitality operation in your commuting area.
2. After calling prospective employers to inquire about job openings, address cover letters and resumes to operations that have open positions you are interested in.

There are pros and cons to both approaches, but generally the second one yields better results. Blanketing area employers with a generic cover letter and resume does little to endear you to any recruiters. Recruiters like to see that you know something about their company and that you can contribute to it. So take the time to target your letters and resumes to the appropriate companies.

Taking a Job Interview

The final step in the process of securing employment in which you will actively participate is the interview. The interview is an arranged meeting between you and your prospective employer, so that he or she may find out more about you. On the basis of this meeting, the interviewer, who will try to learn as much about you as possible, will decide whether you can handle the job and whether you will do the job in the way the company wants it done. As the interviewee, part of your purpose in interviewing is to learn as much as possible about the company and the job you want to fill.

There are three stages to an interview: preparation, the interview itself, and follow up. Let's look at preparation first.

To get ready for your interview, you need to consider these four questions:

1. What do I know about the employer?
2. What am I going to wear?
3. What am I going to say?
4. What do I need to bring with me?

Do your homework on the company that you will be interviewing with. Ask instructors and peers what they know about the company. Perhaps even visit the operation to see what it is like.

As far as dress is concerned, you are going to have to dress considerably differently for an interview than you dress for school. Even if you should apply to a place whose employees do not "dress up," you should still make sure that you look your professional best for the interview. You are better off dressing conservatively, so wear professional business clothing that is clean and well pressed. Be aware that your appearance will count either for or against you, perhaps as much as any other part of your interview.

Once you have figured out what to wear to give an appropriate, business-like impression, consider how you might answer the following interview questions:

*   Tell me about yourself.
*   Why do you want to work here?

- Why did you leave your last job (or why do you want to leave your present job)?
- Why did you apply for this job?
- What is your greatest strength/weakness?
- Where do you expect to be five years from now?
- Why do you feel qualified for this position?
- What kind of people appeal least/most to you as a supervisor?
- What was your last employer's opinion of you?
- Describe your management style.
- What do you consider your biggest work-related accomplishment?
- What two or three things are most important to you in a job?
- What type of atmosphere do you like to work in?
- Describe your personality.
- What motivates/frustrates you?
- Why did you choose this field of work?
- Describe a difficult problem you've had to deal with.
- Rate your job performance on a scale of 1 to 10.
- What area of your skill/development do you want to improve?
- What do you think of your current/last boss?
- Describe a typical day in your current position.
- What did you learn about yourself in previous positions?
- How would your subordinates describe you?
- Summarize your assets.
- What kind of decisions are most difficult for you?
- Where do you want to be working in five years?
- What can you do for us that someone else cannot do?
- Why should I hire you?*

When formulating your answers to these questions, remember that you need to be honest, positive, and complete in your response.

In addition to thinking about how you will answer these questions, you need to jot down questions you will want to ask about the job, such as the following:

- What qualifications are most important in filling this position?
- What are the job duties for this position?
- Who will be my supervisor?
- What are the hours?
- Why is the job open? Is it a new position?
- What kind of training will you offer?
- What kind of advancement is possible?

Put your list of questions, along with extra copies of your resume, a reference list with addresses and phone numbers, copies of reference letters, professional membership cards, and any other pertinent materials into a professional folder that you will take to the interview.

---

*Reprinted with permission from "Taking the Fear Out of Job Interviews" by Linda Eck, *Dietary Manager,* September/October 1994.

Now it is finally show time! Because the interview is the first meeting between you and your prospective employer—and a relatively brief meeting as well—your interviewer will have to base most of her decisions about you on first impressions. The manner in which you introduce yourself to your prospective employer, your personal appearance at the interview, whether you maintain eye-to-eye contact with the interviewer throughout the conversation, the completeness and honesty of your answers to her questions, whether you are on time—these factors will combine to form the interviewer's appraisal of you, both as a person and as a prospective employee.

Make a favorable impression instantly by greeting the interviewer by name, then introducing yourself in a professional, self-confident manner. The person who is conducting the interview will begin to form an opinion of you upon such things as the firmness of your handshake, the clearness of your voice, and whether your demeanor is straightforward. In such a situation, there is really not enough time for the interviewer's impression of you to change significantly. It is therefore very important to get the interview off to a good start.

Now that the interview is moving along, remember these interviewing pointers:

- Look directly at the person who is interviewing you when he is talking.
- Use good body posture. In other words, don't slouch in the chair.
- Don't chew gum.
- Speak clearly, firmly, and not too quickly or slowly.
- Generate friendliness and warmth.
- Be as calm, cool, and collected as you can. Remember that no one is going to bite your head off and the worst that can happen is that you won't be offered a job. Think of the interview as an opportunity to meet some people you may really like.
- Maintain your self-confidence throughout the interview. Don't be modest, but at the same time don't brag!
- Show your enthusiasm both verbally and nonverbally for working for this employer.
- Answer each question completely, directly, and honestly.
- Relate this job to your previous work experiences. Go ahead and tell stories about past jobs that give the interviewer useful and positive information about you as an employee.
- If the interviewer asks your opinion of a former or current employer, speak positively. Never speak negatively about a former or current employer. It serves none of your purposes and will only lower the interviewer's estimation of you. Try to put yourself in the place of the interviewer. If you speak poorly about one employer, what is to prevent you from speaking poorly about another?
- Be a good listener. Do not interrupt the interviewer while she is speaking. Instead of anticipating the interviewer's next questions, concentrate on each question as it is being asked. Being a good listener is an excellent way to build a good relationship with the interviewer.
- Don't save up all your questions for the end of the interview. Ask some of your questions during the interview if the topic is being discussed.
- Although stressful questions are not often used in interviews, they are sometimes employed to see if you can remain graceful under pressure. The inter-

999 Longwood Road
Surfs Up, CA 12345
January 5, 1995

Mr. John White
Director, Food and Beverage
Stately Hotel
111 Stately Avenue
San Francisco, CA 12346

Dear Mr. White,

Thank you for the time and consideration you extended to me during the interview this past Tuesday for the position of Dining Room Server. The interview was interesting as well as informative.

I particularly enjoyed the tour of the dining room. My knowledge of dining room procedures would be an asset in this position.

I am excited about the possibility of working for Stately Hotel and look forward to a positive response.

Sincerely,

Donna Orange
(415) 415-4151

Figure 1.6 Sample Thank-You Letter

viewer may ask you in detail about past job blunders or even try to argue with you and get you to admit you were wrong at some point in the past. If this situation occurs, remember that the interviewer is just testing you, perhaps to see how you might react to an angry customer. Continue to engage in a calm, peaceful discussion, and don't argue.

One last tip has to do with salary. Salary is a difficult area and is not easily approached, but somewhere in the conversation it is going to come up because the dollar is an integral part of any job. Salary is normally discussed toward the end of the interview. Let the interviewer bring it up so you don't give the impression that you are interested only in the money and not the job itself. When you are asked about salary, try to refrain from giving a precise figure. Remember, it will be to your advantage to appear flexible on all counts, including money. This does not mean that you should accept a salary that is inappropriate to your level of employment. It simply means that you should be willing to negotiate a salary level with your prospective employer. By being negotiable about your salary, you will not only be showing your employer that you are interested in the job more than the money but will also be displaying your flexibility as an employee, thus making yourself a more attractive applicant in the interviewer's eyes.

Be sensitive enough to the interview process to be able to tell when the interview is over and it is time to leave. Before the interview is over, be sure to find out from your interviewer what the next step will be. Are you to contact the interviewer, or is the interviewer to contact you? How long will it take for the interviewer to reach a decision? Are you to get in touch with the interviewer if you do not hear from him by a certain date? Should you contact the interviewer by phone or by letter? It is important that you find out from your interviewer how you are supposed to follow up on your interview and to follow her directions precisely.

Be sure to make your closing statement a very positive one. You have gone to the interview expecting to get the job; it is hoped that you have maintained this attitude throughout the interview. Now you want to leave the interviewer with the same positive feelings about you that you have presented to him throughout your meeting, thus indicating that you expect something good to come of the interview.

After an interview, it is a good idea to write a brief note thanking your interviewer for meeting with you. Your courtesy will be appreciated, and your note will further strengthen the already positive impression that you have made upon the company. Also, your thank-you letter will serve as a sort of reminder to your interviewer—she may have interviewed several other applicants for the job and, as a result, your individual characteristics may no longer be as clear as they were during the interview itself. A letter of thanks will do much to restore your interviewer's memory of your performance on the interview, yet it will involve very little time and effort on your part and may very well provide the extra little push that you need to be hired. So make sure to write a brief thank-you letter (see Figure 1.6) and don't delay—write it the same night of the day of your interview.

The more interviews you go on, the more proficient you will become at handling them and the better the interview will go for you. This is one of the best reasons for taking advantage of every opportunity you have to interview with a company. It is especially important for a student who is seeking a part-time or a first full-time job to interview with as many different companies as possible. Going through the interview process a few times will also acquaint you with the jobs available, and this increased familiarity may in turn help you to solidify important career decisions.

1. Know what's expected of you, and what's not expected of you.
2. Ask questions (and not just when you need to find a restroom).
3. Be humble—the person working next to you may have been doing this job for 20 years.
4. Be friendly, or at least look friendly by smiling.
5. Don't be afraid to talk, but also spend lots of time listening.
6. Work hard and absorb as much as you can.
7. Try not to judge other employees—no one asked you to anyway.
8. Offer to do special projects and follow-through on them.
9. Balance your work life with things that you enjoy during your time off.
10. Give it your best shot, that's all any mortal human can do.

Now that you have (we hope!) been offered a job that you want, congratulations are in order, as are some words of advice, so here goes!

Training Goals and Training Guides for Hotels

This chapter includes Training Goals and Training Guides for the following functions in hotels:

- Reservations
- Guest registration and check-out
- Room cleaning
- Laundry
- Room service
- Guest services
- Entertainment and recreation
- Meeting and conferences
- Security
- Accounting and finance

Depending on the nature of your company, some of these functions may be extensive and well developed, such as meetings and conferences in a hotel that primarily serves this market. Other functions may be much less extensive and only involve one or two employees, if that.

Whichever the case, read the Training Goals carefully and fill in the Training Guides as accurately and completely as possible. Don't feel embarrassed to ask questions of employees and supervisors. You'll soon find out that most employees are delighted to tell you about their jobs.

Before starting to use the Training Goals and Training Guides, check with your instructor to see if you should complete the Introductory Worksheet that starts this chapter: Introduction to Hotels. This worksheet is designed to be used initially to help you see the "big picture" of your hotel. Good luck!

Name: _____

Date(s): _____

Department/Location: _____

Source of Information: _____

1. What type of hotel is this?

   _____

   _____

   _____

2. State characteristics of the hotel's clientele. Are they mostly business executives or families on vacation, for example? Include age, income, and profession.

   _____

   _____

   _____

   How long is the average length of stay?

   _____

   _____

3. How long has the hotel been in operation?

   _____

   _____

4. What is the room breakdown of the hotel?

   Total rooms _____

   Total single rooms _____

   Total double rooms _____

   Total suites _____

   Total other types _____

5. Diagram, in organizational-chart form, the departments and jobs in the hotel. **Attach** a chart if available.

6. Is the hotel part of a chain, or is it independently owned?

_____

_____

If independently owned, does the hotel belong to a reservations network?

_____

_____

7. Describe the location of the hotel as well as any unique features.

_____

_____

_____

8. What services are offered at the hotel (such as restaurants, recreation, or meeting facilities)?

_____

_____

_____

_____

_____

_____

9. Is this hotel listed in any guidebooks? If so, how is it rated?

_____

_____

_____

_____

_____

_____

10. How do you rate the accommodations and services offered here?

_____

_____

_____

_____

_____

11. Does the hotel have a formal mission or objective statement? If so, write it below or attach a copy.

_____

_____

_____

_____

_____

_____

_____

12. Is the hotel involved in community activities in the area where you are located?

_____

_____

In what activities has the hotel participated in the past year?

_____

_____

_____

_____

_____

13. How does the operation market its services to potential clientele? Who are the guests it wants to attract?

_____

_____

_____

_____

_____

Does it advertise? Describe where it advertises. **Attach** *a newspaper or magazine ad if possible.*

_____

_____

_____

_____

_____

What are its key advertising themes?

_____

_____

_____

_____

_____

14.

15.

1. Draw an organizational chart and list the duties of reservations personnel.

2. Know room types and locations.

3. Process reservations for individuals and groups.

4. List the reservations policies.

5. Explain how room rates are set and who sets the rates.

6. List statistics reservations generates.

7. _____

8. _____

Name: _____

Date(s): _____

Department/Location: _____

Source of Information: _____

1. Describe how the reservations area is organized and how many employees work here.

_____

_____

_____

_____

_____

_____

_____

_____

2. Sketch the locations of rooms in relation to the front desk.

3. How do reservations come into the hotel (such as by telephone or through a travel agent)?

_____

_____

_____

_____

_____

_____

4. Must a guest leave a deposit on a reserved room or leave a credit card number?

_____

_____

_____

_____

_____

5. What is the daily cutoff time for reservations?

_____

_____

_____

_____

6. How late are reservations held?

_____

_____

7. Under what conditions will the hotel not accept a reservation?

_____

_____

_____

_____

_____

8. How many walk-ins might occur on a typical day?

_____

_____

_____

9. Under what conditions can a guest cancel a reservation without being charged?

_____

_____

_____

_____

_____

10. Does the hotel routinely overbook? By what percentage?

_____

_____

_____

_____

If the hotel is overbooked, how is it decided which guests will not receive rooms?

_____

_____

_____

Does your hotel compensate walked guests? How?

_____

_____

_____

_____

_____

11. How does your hotel deal with no-shows?

_____

_____

_____

_____

_____

_____

12. Describe how room rates are set.

_____

_____

_____

_____

_____

_____

_____

13. Describe the procedure that the reservations clerk follows when making reservations. Be sure to describe the information the clerk needs to make the reservation and how confirmation and/or guarantees are communicated to the requester. **Attach** reservations forms if available.

_____

_____

_____

_____

_____

_____

_____

_____

_____

_____

14. Describe how and when the reservations records are brought to the front office.

_____

_____

_____

_____

15. Describe how the reservationist blocks a group of rooms for organizations or other groups.

_____

_____

_____

_____

16. Does the reservations area initiate forecasts of room sales and occupancy or produce other reservations-related statistics? If so, describe them.

_____

_____

_____

_____

_____

_____

How accurate is the forecasting?

_____

_____

_____

17.

18.

1. Draw an organizational chart and list the duties of front desk personnel.

2. Register and assign a room and room rate to a new guest.

3. Receive and store guests' valuables in a safe.

4. Explain how charges and credits are posted to a guest's folio.

5. Check a guest out of the hotel.

6. _____

7. _____

Name: _____

Date(s): _____

Department/Location: _____

Source of Information: _____

1. Sketch below the hotel lobby, including the front desk.

2. Sketch below the organizational chart for the front desk functions.

3. List the major duties of each job in this area.

_____

_____

_____

_____

_____

_____

_____

_____

_____

_____

_____

_____

_____

4. Describe the type of room rack used at the front desk.

_____

_____

_____

_____

_____

_____

_____

5. When does the expected arrivals list for the day come to the front office?

_____

6. Describe the steps that the front desk clerk takes to register a new guest. **Attach** a copy of your registration form.

_____

_____

_____

_____

_____

_____

_____

_____

_____

_____

What information is given to the guest at this time?

_____

_____

_____

_____

Are guests invited to leave valuables (such as jewelry) in a safe deposit box? If so, describe.

_____

_____

_____

_____

_____

7. Who assigns the room to the guest and determines the room rate?

_____

_____

What are these decisions based on?

_____

_____

_____

_____

_____

_____

When are these decisions made?

_____

_____

_____

_____

When is the clerk allowed to quote a rate lower than the rack rate?

_____

_____

_____

_____

8. Are front office clerks trained to use suggestive selling techniques to sell rooms and to promote other services of the hotel? If so, describe these techniques.

_____

_____

_____

_____

_____

9. What taxes are added to the guest's bill? What percentage is each one?

_____

_____

_____

Are any persons or organizations exempt from paying taxes?

_____

_____

_____

10. What charges are posted to the guest's folio during his or her stay? **Attach** a guest folio if available.

_____

_____

_____

_____

How are charges posted onto the proper guest folio? Who posts the charges?

_____

_____

_____

_____

_____

What types of credits might be posted to a guest's folio?

_____

_____

_____

Where are guests' folios kept?

_____

_____

_____

11. How much credit is typically extended to a guest?

_____

_____

_____

_____

12. Describe the process in which a guest checks out.

_____

_____

_____

_____

_____

_____

_____

_____

What is the job title of the person who normally handles check-out?

_____

_____

Is express or rapid check-out service available? If so, how does it work?

_____

_____

_____

_____

13.

14.

1. Draw an organizational chart and list the duties of housekeeping personnel.

2. Clean an occupied and unoccupied room.

3. Communicate housekeeping needs to room attendants and communicate room readiness to the front desk.

4. Know the supplies that room attendants need to do their job and how to restock their cart.

5. Know procedures for room attendants to follow when lost or misplaced property is found.

6. Inspect cleaned rooms.

7. Inspect public areas of the hotel for cleanliness.

8. Determine daily housekeeping costs per occupied room.

9. _____

10. _____

Name: _____

Date(s): _____

Department/Location: _____

Source of Information: _____

1. Sketch below the organizational chart for the housekeeping area.

2. How many rooms per day are room attendants expected to clean?

_____

_____

_____

_____

3. How is the room attendant informed of his or her job assignments for the day? **Attach** a form if available.

_____

_____

_____

_____

How does the room attendant stay in contact with the department and/or the front office for room status changes and to report when rooms are ready?

_____

_____

_____

_____

4. How much time is allowed to clean an occupied room?

_____

_____

_____

How much time is allowed to clean an unoccupied room?

_____

_____

_____

5. What supplies are on the attendant's cart?

_____

_____

_____

When is the cart restocked?

_____

_____

What control is there over the amount of linens on the cart?

_____

_____

_____

_____

How are the cleaning supplies and expendable items issued to room attendants?

_____

_____

_____

_____

6. List the steps taken to clean an unoccupied room. If there is a form of some type that describes the steps, **attach** the form.

_____

_____

_____

_____

_____

_____

_____

7. What is the difference in cleaning procedures and duties for an occupied and an unoccupied room?

_____

_____

_____

_____

_____

_____

8. What procedures are followed when a room attendant finds a room that has been torn up and has broken furniture, spills, and so forth?

_____

_____

_____

_____

_____

9. What is the procedure for returning or turning in the lost or misplaced property of a guest?

_____

_____

_____

_____

_____

10. Who inspects the rooms after the room attendant has cleaned them?

_____

_____

Do all rooms get inspected? If not, how are rooms selected for inspection?

_____

_____

_____

_____

11. How often and when are public areas in the hotel, such as halls, lobby, and restrooms, cleaned, and who cleans them?

_____

_____

_____

_____

12. Are there any specific architectural features or building materials that cause housekeeping problems?

_____

_____

_____

_____

How can they be worked around in existing buildings?

_____

_____

_____

_____

13. How often is a major cleaning of rooms done? What is done and who does the cleaning?

_____

_____

_____

_____

_____

_____

_____

_____

14. What special projects are undertaken by room attendants during low-occupancy periods?

_____

_____

_____

_____

_____

15. What are the most common accidents, or types of accidents, in the housekeeping department?

_____

_____

_____

_____

_____

What is being done to reduce the number of accidents?

_____

_____

_____

_____

16. Are linens rented, or does the hotel run its own laundry?

_____

_____

17. Discuss with a housekeeping manager or supervisor the advantages and disadvantages of linen rental and a self-operated laundry.

_____

_____

_____

_____

_____

_____

_____
_____
_____

18. How often does the head of housekeeping meet with housekeeping managers and supervisors?

_____
_____
_____

What topics are most often discussed?

_____
_____
_____
_____
_____
_____

19. Can you determine daily laundry, cleaning supply, expendable, and labor costs for an occupied room? List the costs.

_____
_____
_____
_____
_____
_____

20. How does the head of housekeeping stay informed and current on new products and procedures in the housekeeping field?

_____
_____
_____
_____
_____
_____

21. What procedure is used to select the cleaning supplies and expendables to be purchased by the department?

_____

_____

_____

_____

_____

21.

22.

1. Draw an organizational chart and list the duties of laundry personnel.

2. Count linen out and in.

3. Determine the amount of linen required for the operation.

4. Take a linen inventory.

5. Describe controls to prevent linen theft.

6. _____

7. _____

Name: _____

Date(s): _____

Department/Location: _____

Source of Information: _____

1. List the duties of each laundry job.

_____

_____

_____

_____

_____

_____

_____

_____

_____

_____

_____

_____

_____

_____

2. Does the laundry wash linens used by the dining rooms and catering?

_____

Does the laundry launder employee uniforms?

_____

3. Give the procedure used for counting linen out and in when a laundry or rental service is used.

_____

_____

_____

_____

_____

_____

_____

4. How many linen items are required for each bed to ensure an ample supply? Take into consideration the linen in circulation and what linen will be needed for days when the laundry is closed.

_____

_____

_____

_____

_____

5. How often is a physical inventory of the hotel linen taken?

_____

What is the inventory procedure, and who takes the inventory?

_____

_____

_____

_____

_____

6. What is the approximate dollar value of linen loss per year from employee and/or guest theft? Provide a breakdown of the items lost if one is available.

_____

_____

_____

_____

7. List the controls used to keep down linen theft and loss.

_____

_____

_____

_____

_____

8.

9.

1. Take room service orders accurately and completely.

2. Communicate room service orders to the appropriate stations.

3. Fill and deliver room service orders according to hotel procedures.

4. _____

5. _____

Name: _____

Date(s): _____

Department/Location: _____

Source of Information: _____

1. Does the hotel offer room service?

_____

2. **Attach** a room service menu or list the foods offered.

_____

_____

_____

_____

_____

_____

_____

_____

_____

3. During which hours is room service offered?

_____

_____

_____

4. Which menu items are the most popular?

_____

_____

_____

_____

5. Compare the room service prices to the hotel dining room prices.

_____

_____

6. What is the average delivery time for room service?

_____

_____

7. How is hot food kept hot during delivery?

_____

_____

_____

8. Describe the process of taking room service orders on the phone and filling and delivering the order. **Attach** an order form if available.

_____

_____

_____

_____

_____

_____

_____

_____

_____

9. What do you do if you deliver an order and there is a "PLEASE DON'T DISTURB" sign on the door?

_____

_____

_____

10.

11.

1. List the types of services available to guests.

2. Take luggage to guests' rooms.

3. Answer guests' questions concerning hotel and local services.

4. Be knowledgeable about the telephone answering system and how guests receive phone messages.

5. _____

6. _____

Name: _____

Date(s): _____

Department/Location: _____

Source of Information: _____

1. Which of the following guest services are available at your hotel?

   Banking services   _____

   Bell staff   _____

   Concierge   _____

   Door greeters   _____

   Dry cleaning   _____

   Garage and parking attendants   _____

   Guest laundry   _____

   Hair styling   _____

   Health facilities   _____

   Office services   _____

2. Describe the principal duties of each guest services job available at your hotel. Include to whom each job reports.

_____

_____

_____

_____

_____

_____

_____

_____

_____

_____

_____

3. What are the most frequently asked questions that the concierge (or front desk clerk if there is no concierge) must answer?

_____

_____

_____

_____

_____

4. Describe how guests get telephone messages.

_____

_____

_____

_____

_____

5. Describe how guests can receive and send fax transmissions, regular mail and package service, and overnight mail and package service.

_____

_____

_____

_____

_____

_____

_____

_____

6. Describe the telephone operator's job. Does it include making wake-up calls and paging?

_____

_____

_____

_____

_____

_____

_____

7. How are telephone charge amounts determined?

_____

_____

_____

_____

_____

_____

8.

9.

1. Be knowledgeable about how to obtain appropriate entertainment, entertainment costs, and promoting entertainment events.

2. Discuss how the payment of royalty (copyright) fees for music are made.

3. Explain the rules for using recreation facilities.

4. _____

5. _____

Name: _____

Date(s): _____

Department/Location: _____

Source of Information: _____

1. Does the hotel book live entertainment?

_____

_____

What types of live entertainment are hired?

_____

_____

_____

_____

Where does the live entertainment usually perform?

_____

_____

What is the length of the typical engagement?

_____

_____

How often is live entertainment brought in?

_____

_____

_____

2. How does the hotel locate suitable entertainment?

_____

_____

_____

_____

3. How does the hotel advertise and promote the entertainment? **Attach** any advertising if available.

_____

_____

_____

_____

4. To whom are the various types of entertainment marketed?

_____

_____

_____

_____

5. Which type of entertainment seems the most popular and successful?

_____

_____

_____

6. What is the size of the entertainment budget?

_____

_____

_____

Compare this figure to total food and beverage sales.

_____

_____

_____

7. Does the hotel pay any royalty (copyright) fees for music offered? To whom are the fees paid, and how much are they?

_____

_____

_____

_____

_____

8. List all recreational facilities available, such as a pool or fitness room.

_____

_____

_____

_____

_____

9. How large is the swimming pool, if available? Is it indoor or outdoor?

_____

_____

_____

_____

10. What are the hours of the pool? Is a lifeguard always on duty?

_____

_____

_____

_____

_____

_____

11. List the pool rules or **attach** a copy.

_____

_____

_____

_____

_____

_____

_____

_____

12. Who is responsible for maintaining the cleanliness and safety of the pool's water?

_____

_____

What does this person have to do to keep the pool water clean?

_____

_____

_____

_____

_____

13. What equipment is available in the fitness room (if available)?

_____

_____

_____

_____

_____

_____

14. What facilities are available, either in the hotel or outside, for joggers?

_____

_____

_____

15. Are there any recreational facilities designed for children, such as a playground? If so, describe them.

_____

_____

_____

16.

17.

1. Answer questions about types of meeting rooms and other facilities available.

2. Describe the steps in selling meetings and conferences.

3. Explain how sales/marketing coordinates with all hotel departments after booking.

4. Maintain appropriate sales/marketing records.

5. _____

6. _____

Name: _____

Date(s): _____

Department/Location: _____

Source of Information: _____

1. What advertising is done to promote meetings and conferences at your hotel? **Attach** a sample if available.

_____

_____

_____

_____

_____

Where is the advertising done?

_____

_____

_____

_____

_____

2. What is the average number of meetings and conferences held at your hotel each week?

_____

_____

What percentage of the total hotel income is derived from meetings and conferences?

_____

_____

3. What facilities are available at your hotel for meetings and conferences? Include the size of meeting rooms and exhibit rooms as well as types of audiovisual equipment available.

_____

_____

_____

_____

_____

_____

_____

_____

Which hotel services, such as a fitness room, or community attractions are popular with meeting/conference attendees?

_____

_____

_____

_____

_____

4. What is the largest meeting that your hotel can accommodate?

_____

_____

_____

_____

5. Sketch below the organizational structure of the sales or marketing department responsible for booking meetings and conferences.

6. Give a brief list of the duties of each job in the sales or marketing department.

_____

_____

_____

_____

_____

_____

_____

_____

_____

_____

7. Describe the role of a meeting planner and how the meeting planner interacts with the sales managers.

_____

_____

_____

_____

_____

_____

_____

8. Ask a sales manager to describe the steps in the sales process.

_____

_____

_____

_____

_____

_____

_____

_____

_____

_____

9. How does the sales manager coordinate a meeting or conference with all departments in the hotel after booking it?

_____

_____

_____

_____

_____

_____

10. Who makes the final check of meeting rooms to make sure they are satisfactory?

_____

_____

What happens if a meeting room is not set up correctly?

_____

_____

_____

11. What is the typical room rate paid by meeting/conference attendees?

_____

_____

Who decides what the room rate is, and what is it based on?

_____

_____

_____

_____

_____

12. Are check-in and check-out procedures made easier for meeting/conference attendees? If so, describe how.

_____

_____

_____

_____

_____

_____

13. What is the policy of the hotel on complimentary services, rooms, food, and beverages for any meeting/conference attendees?

_____

_____

_____

_____

_____

_____

_____

14. What records are maintained by the sales/marketing department, and how are they used to book or sell business?

_____

_____

_____

_____

_____

_____

15.

16.

1. Identify security measures taken to protect employees.

2. Identify security measures taken to protect guests.

3. Identify security measures taken to prevent thefts by employees.

4. Identify security measures taken to prevent theft by guests.

5. Take appropriate actions when guests walk out without paying and so on.

6. Report security incidents appropriately.

7. Help in security investigations.

8. _____

9. _____

Name: _____

Date(s): _____

Department/Location: _____

Source of Information: _____

1. Sketch below the organizational structure of the department in charge of hotel security.

2. Make a list of the major duties of each job in security.

_____

_____

_____

_____

_____

_____

_____

_____

_____

_____

_____

_____

_____

3. What does security do to supervise entry and exit of hotel employees?

_____

_____

_____

_____

Do hotel employees have to enter and exit the hotel through designated doors? If so, explain why.

_____

_____

_____

_____

Do hotel employees wear identification badges? If so, describe how the badges look and how they are used.

_____

_____

_____

_____

Are employees' packages and bags checked during entry and exit from the building? If so, why?

_____

_____

_____

Are there any situations in which an employee may leave with something that belongs to the hotel? If so, explain.

_____

_____

_____

_____

4. List five measures established to protect guests, including access to their rooms.

_____

_____

_____

_____

_____

_____

5. Describe the process by which security investigates reported incidents, such as theft.

_____

_____

_____

_____

_____

_____

_____

_____

_____

6. Discuss how incidents are reported (such as by phone or on special forms). **Attach** the forms if available.

_____

_____

_____

_____

_____

_____

7. What is the most frequent incident that security becomes involved in?

_____

_____

8. Which incident has the greatest negative impact in terms of costs and lowered productivity?

_____

_____

_____

9. In cases of theft, which items are most often taken?

_____

_____

_____

10. Does security use any sensors, alarm systems, or special surveillance equipment to check for internal and external theft? If so, describe them.

_____

_____

_____

_____

_____

_____

11. Are all employees who handle cash, checks, and credit cards trained to recognize forgery and counterfeit money, money orders, travelers' checks, and drivers' licenses, and how to deal with quick-change artists?

_____

_____

_____

_____

_____

12. List 10 other measures taken to prevent cash theft by robbers or by employees.

_____

_____

_____

_____

_____

_____

_____

_____

_____

_____

13. Give three reasons that employees steal.

_____

_____

_____

_____

_____

_____

14. If an employee sees a guest taking an item that belongs to the establishment, what is the employee supposed to do?

_____

_____

_____

_____

_____

15. If an employee sees a guest leaving without paying, what is the employee supposed to do?

_____

_____

_____

_____

_____

16. Describe the areas in which security actively trains employees on security procedures.

_____

_____

_____

_____

_____

_____

_____

_____

17. Is security involved in any aspect of hiring any employee or employees for certain jobs? If so, describe the involvement.

_____

_____

_____

_____

_____

_____

18. List the areas for which the hotel has emergency plans. **Attach** a sample plan if available.

_____

_____

_____

_____

_____

_____

_____

Describe what should happen in the event of a fire.

_____

_____

_____

_____

_____

_____

19. What type of public liability insurance is carried by the operation?

_____

_____

_____

_____

How much coverage is provided?

_____

_____

_____

_____

What are the premiums?

_____

_____

20.

21.

1. Explain how accounting and finance are organized and the major duties of each job.

2. Interpret income statements.

3. Calculate average rate per occupied room, average income per guest, and other averages as needed.

4. Calculate profitability, operating, and other ratios as needed.

5. Calculate standard food or beverage costs.

6. Identify internal controls.

7. Explain how operating budgets are prepared.

8. _____

9. _____

Name: _____

Date(s): _____

Department/Location: _____

Source of Information: _____

1. Does your company use an outside accounting firm for bookkeeping and accounting services?

    _____

    If so, what services do they provide and at what cost?

    _____

    _____

    _____

    _____

    _____

    _____

2. If an outside firm is not used, sketch below the organizational structure of the area in charge of accounting and finance.

3. Make a list of the major duties of each job in accounting and finance.

    _____

    _____

    _____

_____

_____

_____

_____

_____

_____

_____

_____

4. Who prepares income statements (or profit and loss statements) for the hotel? How often are they prepared?

_____

_____

_____

What important information does the income statement provide to the company?

_____

_____

_____

_____

_____

_____

_____

_____

_____

5. List the ratios the company generates and looks at, such as operating and profitability ratios.

_____

_____

_____

_____

_____

_____

_____

_____

_____

_____

6. List the averages the company generates and looks at, such as the average rate per occupied room.

_____

_____

_____

_____

_____

_____

_____

_____

_____

7. Explain how the company calculates a standard food or beverage cost.

_____

_____

_____

_____

_____

_____

8. Identify 10 internal controls.

_____

_____

_____

_____

_____

_____

_____

_____

_____

_____

9. Describe the major steps in putting together an operating budget for the hotel.

_____

_____

_____

_____

_____

_____

_____

_____

_____

_____

_____

_____

_____

10. What time period does the budget cover?

_____

_____

11. Is the operating budget fixed or flexible?

_____

_____

12. What categories do hotel managers budget for?

_____

_____

_____

_____

_____

_____

_____

_____

_____

_____

_____

_____

13. Once an operating budget is determined, how is it used?

_____

_____

_____

_____

_____

_____

_____

14. What are the key elements in a capital budget?

_____

_____

_____

_____

_____

15.

16.

Training Goals and Training Guides for Restaurants and Foodservices

This chapter includes Training Goals and Training Guides for the following functions in restaurants and foodservices:

- Managing kitchen operations
- Purchasing
- Receiving, storeroom, and inventory
- Cooking
- Managing dining room service
- Serving tables
- Counter service
- Cafeteria
- Cash handling
- Beverage service
- Wine service
- Catering
- Vending
- Sanitation and safety
- Security
- Accounting and finance

Depending on the nature of your company, some of these functions may be extensive and well developed, such as wine service in a white tablecloth restaurant. Other functions may be much less extensive and only involve one or two employees, if that.

Whichever the case, read the Training Goals carefully and fill in the Training Guides as accurately and completely as possible. Don't feel embarrassed to ask questions of employees and supervisors. You'll soon find out that most employees are delighted to tell you about their jobs.

Before starting to use the Training Goals and Training Guides, check with your instructor to see if you should complete the Introductory Worksheet that starts this chapter: Introduction to Restaurants and Foodservices. This worksheet is designed to be used initially to help you see the "big picture" of your restaurant or foodservice. Good luck!

Name: _____

Date(s): _____

Department/Location: _____

Source of Information: _____

1. Does the operation have a formal mission or objective statement? If so, write it below or **attach** a copy.

_____

_____

_____

_____

_____

_____

2. Into how many departments or functional areas (such as dining room or food production) is the operation organized? **Attach** an organizational chart if available. What are the departments or functional areas, and how many employees are in each department?

_____

_____

_____

_____

What is the name of the person in charge of each department?

_____

_____

_____

_____

_____

_____

_____

_____

_____

_____

3. What are the job titles (such as cook) found in the operation? List them.

_____

_____

_____

_____

_____

_____

_____

4. What are the operation's hours of operation?

_____

_____

_____

5. Who schedules the employees? Is there one master schedule, or does each area/department schedule its own employees?

_____

_____

_____

_____

6. Which, if any, employees wear uniforms? Describe the uniforms.

_____

_____

_____

_____

Who pays for and launders the uniforms?

_____

_____

_____

_____

7. Are meals provided to employees on duty? If so, which foods are employees allowed to eat?

_____

_____

_____

_____

_____

8. Sketch out below the various rooms that make up the operation. Indicate how each room is used, such as for cooking, serving guests, and so on. Use arrows to indicate the flow of food from the receiving of raw goods to the serving of a finished product.

9. Describe the type of menu(s) available. Is each menu static, cycle, or single use? Be sure to mention menus for different areas, such as cafeteria and patient tray service. **Attach** a menu.

_____

_____

_____

_____

Which meals are available?

_____

_____

_____

_____

How wide is the variety of choices for each meal?

_____

_____

_____

_____

Who is responsible for planning and revising menus?

_____

_____

_____

_____

How often are the menus revised?

_____

_____

_____

_____

How are menu prices determined?

_____

_____

_____

_____

10. How are food and beverages served to the guests (table service, counter service, cafeteria service, and so on)?

_____

_____

_____

_____

11. What is the total number of seats?

_____

12. How many meals are served daily?

_____

13. How is customer satisfaction measured? Describe in adequate detail. **Attach** any forms used.

_____

_____

_____

_____

_____

14. Is the organization involved in community activities in the area where you are located?

_____

_____

In what activities has the organization participated in the past year?

_____

_____

_____

_____

_____

15. How does the operation market its services to potential clientele? Describe this marketing in adequate detail.

_____

_____

_____

_____

_____

_____

_____

16.

17.

1. Draw the organizational structure of all personnel, such as the Executive Chef, involved in managing the kitchen operation.

2. Describe the duties of kitchen managers, including assistant managers.

3. Write and post service personnel schedule in designated location.

4. Verify and total hours worked/nonworked for payroll.

5. Complete, or assist in completing, performance appraisals for kitchen personnel.

6. Train new kitchen personnel.

7. Run a meeting of kitchen personnel to review policies and procedures, discuss new policies and procedures, receive employee input, etc.

8. _____

9. _____

Name: _____

Date(s): _____

Department/Location: _____

Source of Information: _____

1. Draw the organizational structure of all personnel, such as the Executive Chef, involved in managing the kitchen operation.

2. Who (name and title) is responsible for managing:

   - purchasing?

   _____

   - receiving and storing?

   _____

   - cooking?

   _____

   - interfacing between cooking and service personnel?

   _____

   - sanitation?

   _____

   - employee safety

   _____

3. List each job involved in managing kitchen operations and include the job's main duties.

_____

_____

_____

_____

_____

_____

_____

_____

_____

4. What are the qualifications for each of the jobs listed above?

_____

_____

_____

_____

_____

_____

_____

_____

_____

_____

5. What days and hours do kitchen managers work?

_____

_____

_____

_____

6. Are any of the current managers people who worked their way up from an entry-level job? If so, identify by title.

_____

_____

_____

7. Ask one of the managers what he or she likes most and least about the job. Describe below.

_____

_____

_____

_____

_____

_____

8. How involved are kitchen managers in human resources functions such as staffing, orientation, training, payroll and evaluation? Describe.

_____

_____

_____

_____

_____

_____

_____

_____

_____

_____

9. Do any of the kitchen managers hold regular meetings with kitchen staff? Describe.

_____

_____

_____

10.

11.

1. Determine specifications and quality standards.

2. Determine purchase quantities.

3. Select suppliers and maintain good supplier relations.

4. Conduct negotiations with suppliers to get the best value.

5. Conduct research activities such as value analysis, forecasting, make-or-buy analysis, and vendor visits.

6. Maintain an optimal level of inventory items so as to minimize investment and increase freshness.

7. Develop and utilize adequate recordkeeping control documents.

8. _____

9. _____

Name: _____

Date(s): _____

Department/Location: _____

Source of Information: _____

1. Who in the operation is responsible for purchasing? List each person's name, title, and buying responsibility.

   _____

   _____

   _____

   _____

   _____

2. Does the buyer (or buyers) set hours and days when he will see salespeople? If yes, what are the days and hours, and why is this policy used?

   _____

   _____

   _____

   _____

3. When salespeople come in, what types of information does the buyer ask for?

   _____

   _____

   _____

   _____

   _____

4. Ask the buyer what criteria are used to select vendors and list the criteria below.

   _____

   _____

   _____

   _____

5. Does the buyer actively negotiate prices with vendors? If yes, describe the negotiations.

_____

_____

_____

_____

6. Are orders placed by telephone, by formal bids, standing orders, or other methods?

_____

_____

_____

_____

_____

_____

_____

Is competitive bidding required for any products you purchase? If so, list the products.

_____

_____

_____

_____

7. What is the dollar amount for annual food purchases?

_____

_____

_____

What percent of this total is spent on the various categories of foods, such as meat and produce.

_____

_____

_____

_____

_____

_____

_____

_____

_____

8. Does the buyer visit vendors at their place of business? If so, what is the purpose of these visits, and how often are they done?

_____

_____

_____

_____

_____

9. What type of testing is done on a new or different product before it is accepted or rejected by the buyer? Who does the testing?

_____

_____

_____

_____

10. Describe any other research activities undertaken by the buyer, such as make-or-buy analysis.

_____

_____

_____

_____

_____

11. When the buyer is looking for a new product or something new for the menu, where does she go for information?

_____

_____

_____

12. Does the operation purchase mostly raw foods or prepared/convenience foods?

_____

_____

13. Other than a labor factor, what does your manager see as an advantage of using convenience foods? As a disadvantage?

_____

_____

_____

_____

14. Does your operation receive any government commodity foods? If yes, explain the process of ordering these foods, and give examples of the types of foods available.

_____

_____

_____

_____

_____

_____

15. Is the person responsible for ordering also responsible for receiving, issuing, and taking inventory? If yes, what is your opinion of this system?

_____

_____

_____

_____

16. How is the receiving clerk informed of what has been ordered? **Attach** any forms used if available.

_____

_____

_____

17.

18.

—

Training Guide—Purchasing Meat, Poultry, and Seafood

Name: _____

Date(s): _____

Department/Location: _____

Source of Information: _____

1. Are written specifications used for purchasing meat, poultry, and seafood? Do the specifications include grades? Give an example of a specification for one meat, one poultry, and one seafood product, or **attach** samples.

   _____

   _____

   _____

   _____

   _____

   _____

2. What are the procedures for purchasing meat, poultry, and seafood? Are they formal or informal?

   _____

   _____

   _____

   _____

3. Who determines how much to order? How are these amounts determined?

   _____

   _____

   _____

   _____

4. Are purchase orders or requisitions issued for all or for some items to be purchased?

   _____

   _____

   _____

   _____

5. How often are meat, poultry, and seafood ordered and delivered?

_____

_____

_____

_____

6. Does the operation purchase primal or fabricated cuts of meat? What are the most frequently purchased cuts?

_____

_____

_____

_____

_____

7. What is the composition of the ground beef that you use?

_____

_____

8. Does the operation purchase any portion-control meats, such as four-ounce hamburger patties or six-ounce chicken breasts?

_____

If so, describe. How much tolerance is given the vendor on the actual weight per item?

_____

_____

_____

_____

_____

9. What forms of poultry (such as whole or breasts) are purchased? Do they come in fresh or frozen?

_____

_____

_____

_____

_____

10. What forms of seafood (such as whole or fillets) are purchased? Do they come in fresh or frozen? Why is fresh or frozen preferred?

_____

_____

_____

_____

11. Does the buyer purchase any convenience entrees? If so, describe them.

_____

_____

_____

12.

13.

Name: _____

Date(s): _____

Department/Location: _____

Source of Information: _____

1. Are written specifications used for purchasing produce? Do the specifications include grades? Give an example of a specification for one fruit and one vegetable, or **attach** samples.

   _____
   _____
   _____
   _____
   _____
   _____

2. What are the procedures for purchasing produce? Are they formal or informal?

   _____
   _____
   _____
   _____
   _____

3. Who determines how much to order? How are these amounts determined?

   _____
   _____
   _____
   _____

4. Are purchase orders or requisitions issued for all or for some items to be purchased?

   _____
   _____
   _____
   _____

5. How often is produce ordered and delivered?

_____

_____

_____

_____

6. Does the operation purchase any prewashed, prepeeled, or precut produce, such as peeled potatoes or carrot sticks?

_____

_____

_____

_____

7. Where can the buyer read about the availability, quality, and pricing of fresh produce?

_____

_____

_____

8.

9.

Name: _____

Date(s): _____

Department/Location: _____

Source of Information: _____

1. Are written specifications used for purchasing milk, dairy products, and eggs? Do most specifications include grades? Give an example of a specification for fluid milk, American cheese, margarine, and fresh eggs, or **attach** samples.

   _____

   _____

   _____

   _____

   _____

   _____

   _____

2. What are the procedures for purchasing milk, dairy products, and eggs? Are they formal or informal?

   _____

   _____

   _____

   _____

   _____

3. Who determines how much to order? How are these amounts determined?

   _____

   _____

   _____

   _____

   _____

4. Are purchase orders or requisitions issued for all or for some items to be purchased?

_____

_____

_____

_____

Are any standing orders used? For which products?

_____

_____

_____

5. How often is milk ordered and delivered? Dairy products? Eggs?

_____

_____

_____

_____

_____

6. Does the operation purchase any processed egg products? If so, describe.

_____

_____

_____

_____

7.

8.

Name: _____

Date(s): _____

Department/Location: _____

Source of Information: _____

1. Does the operation purchase a variety of fresh baked goods such as cakes, pies, and muffins?

   _____

   _____

   _____

   _____

2. Are written specifications used for purchasing fresh baked goods? Give an example of a specification for white bread, or **attach** a sample.

   _____

   _____

   _____

   _____

   _____

   _____

   _____

   _____

3. What are the procedures for purchasing baked goods? Are they formal or informal? Are standing orders used? **Attach** a standing order form if possible.

   _____

   _____

   _____

   _____

   _____

   _____

   _____

4. Who determines how much to order? How are these amounts determined?

_____

_____

_____

_____

_____

_____

5. How often are baked goods ordered and delivered?

_____

_____

_____

_____

6.

7.

Name: _____

Date(s): _____

Department/Location: _____

Source of Information: _____

1. Are written specifications used for purchasing dry goods and frozen foods? Do most specifications indicate either a grade or a brand name? Give an example of a specification for one canned good and one frozen food, or **attach** samples.

   _____

   _____

   _____

   _____

   _____

   _____

   _____

   _____

2. What are the procedures for purchasing dry goods and frozen foods? Are they formal or informal?

   _____

   _____

   _____

   _____

   _____

   _____

3. Who determines how much to order? How are these amounts determined?

   _____

   _____

   _____

   _____

   _____

4. Are purchase orders or requisitions issued for all or for some items to be purchased?

_____

_____

_____

_____

5. How often are dry goods and frozen foods ordered and delivered?

_____

_____

_____

_____

6. Does the operation purchase many convenience foods? List at least five convenience foods it buys.

_____

_____

_____

_____

7.

8.

1. Receive all incoming goods by carefully checking amount, quality, and price.

2. Maintain receiving records.

3. Put incoming goods into storage using the first-in–first-out system.

4. Maintain clean and orderly storeroom areas that meet all sanitation guidelines.

5. Issue material from stock.

6. Survey products on hand for lack of movement or overstocking.

7. Take a physical inventory and cost it out.

8. Prepare a needs list for the person who does the ordering.

9. Maintain perpetual inventory records.

10. _____

11. _____

Name: _____

Date(s): _____

Department/Location: _____

Source of Information: _____

1. What are the steps in the receiving process using the operation's policies and procedures?

   _____

   _____

   _____

   _____

   _____

2. Are these steps followed in practice? If not, explain why not.

   _____

   _____

   _____

3. Once a shipment is accepted and the invoice is signed, what happens to the invoice?

   _____

   _____

   _____

4. Does the receiving clerk maintain a daily record of items received? If so, what information is recorded? **Attach** a daily receiving report if available.

   _____

   _____

   _____

5. How quickly are incoming goods put into storage?

   _____

   _____

   _____

   _____

6. Are there any special guidelines for putting goods away, such as marking the date on canned goods? If so, describe them.

_____

_____

_____

_____

7. From a sanitation point of view, are potentially hazardous foods properly received and stored? If not, describe what could be done to improve the situation.

_____

_____

_____

_____

8.

9.

Name: _____

Date(s): _____

Department/Location: _____

Source of Information: _____

1. Describe the size of the storeroom facilities available for dry goods and refrigerated and frozen foods. Is there adequate room for storing foods?

   _____

   _____

   _____

   _____

2. At what temperatures are the different storage areas maintained? Are these temperatures regularly checked and recorded? If so, by whom and how often? **Attach** any temperature log forms if available.

   _____

   _____

   _____

   _____

   _____

3. Are the storage area temperatures within generally accepted sanitation standards?

   _____

   _____

   _____

   _____

4. Who is responsible for cleaning the storage areas? Is there a cleaning schedule for this area? If so, describe it. **Attach** a cleaning schedule if available.

   _____

   _____

   _____

   _____

5. Are storage areas free from pipes, ventilation ducts, and water lines? If not, are these lines causing any problems?

_____

_____

6. Are storage areas properly protected from insects and rodents? If not, describe the problem and how it can be corrected.

_____

_____

_____

7. Is food kept six inches away from walls and at least six inches above the floor?

_____

_____

_____

8. Is food being rotated properly? Describe the system used.

_____

_____

9. Are foods issued from stock to employees according to a specific procedure? Describe the procedure.

_____

_____

_____

10.

11.

Name: _____

Date(s): _____

Department/Location: _____

Source of Information: _____

1. How often is a physical inventory taken?

   _____

2. Why is a physical inventory taken?

   _____

   _____

3. Are all goods, both food and nonfood, inventoried at the same time? Note any items that are inventoried at different intervals.

   _____

   _____

4. What is the procedure for taking physical inventory? **Attach** a sample inventory form if available.

   _____

   _____

   _____

5. Who is responsible for taking the physical inventory?

   _____

   _____

6. Who is responsible for costing and extending prices of the inventory?

   _____

7. When the inventory is costed, how is the unit price determined?

   _____

   _____

   _____

   _____

8. Are any perpetual inventory systems maintained? If so, describe them.

_____

_____

_____

9.

10.

1. Follow recipes to meet quality and quantity standards.

2. Adjust recipes accurately.

3. Use and follow production sheets.

4. Operate cooking and food-processing equipment and hand tools appropriately and safely.

5. Taste-test and evaluate foods before they are served.

6. Portion foods into standard portions.

7. Garnish food in an appealing manner.

8. Handle food in a sanitary and safe manner.

9. _____

10. _____

Name: _____

Date(s): _____

Department/Location: _____

Source of Information: _____

1. Who prepares soups and sauces in your operation? What are the scheduled hours for this cook (or cooks)?

   _____

   _____

   _____

2. What is the most popular soup?

   _____

   Is there a soup that is the signature soup, or specialty, of the house?

   _____

3. Does your operation use convenience bases, or do you make your own stocks? What types of bases are used or stocks prepared?

   _____

   _____

   _____

4. Are soups made from scratch, and/or are convenience products used (such as frozen, canned, or dried soups)?

   _____

   _____

   _____

5. List three soups that you make, and describe briefly how each is prepared.

   _____

   _____

   _____

   _____

   _____

6. Does the cook preparing soups also cut up all the vegetables needed, or are those supplied by another kitchen area?

_____

_____

_____

7. What are the serving sizes for soup? What dishes are used?

_____

_____

_____

8. How are soups garnished?

_____

_____

_____

_____

9. List the types of sauces that are prepared, and note if each is a mother sauce or a small sauce and what dish the sauce is used with.

_____

_____

_____

_____

_____

_____

_____

_____

10. Which sauces are the most popular?

_____

_____

11. Are any ready-to-heat sauces (either frozen or canned) used? If so, list them.

_____

_____

_____

_____

12. Are there any sauces, such as a coulis, that are meant to be lighter and more nutritious than traditional sauces? List them.

_____

_____

_____

13. How are soups and sauces held for service to maintain flavor and quality?

_____

_____

_____

14. List the cooking equipment and tools used in this area.

_____

_____

_____

_____

_____

15. Describe the production sheet used in this area. How are production needs forecast? **Attach** a production sheet if available.

_____

_____

_____

_____

16. Sketch out the area in which the soup and sauce cook works. Indicate all equipment and personnel. Use arrows to show work flow.

16.

17.

Name: _____

Date(s): _____

Department/Location: _____

Source of Information: _____

1. Who prepares the entrées in your operation? What are the scheduled hours for this cook(or cooks)?

_____

_____

_____

2. What is the most popular entrée?

_____

Is there an entrée that is considered the specialty of the house? If so, name it.

_____

3. List the cooking equipment used in this area.

_____

_____

_____

_____

_____

_____

_____

4. What menu items are prepared by grilling?

_____

_____

_____

_____

_____

_____

in the oven?

_____

_____

_____

_____

_____

using the broiler?

_____

_____

_____

_____

_____

by sautéing?

_____

_____

_____

_____

_____

by deep-fat frying?

_____

_____

_____

_____

_____

5. Describe how the broiler station is set up.

_____

_____

_____

_____

Describe how the sauté station is set up.

_____

_____

_____

_____

_____

6. Are any entrées marinated? If so, name them.

_____

_____

7. Describe how entrées are seasoned (before cooking, after cooking, with salt, with fresh herbs, and so on).

_____

_____

_____

_____

_____

_____

8. How does the cook determine doneness of meats, poultry, and seafood?

_____

_____

_____

_____

9. List portion sizes for five popular entrées.

_____

_____

_____

_____

_____

10. How is the entrée arranged on the plate, and how is it garnished?

_____

_____

_____

_____

11. How are entrée orders communicated to the cook?

_____

_____

_____

_____

12. How much time is allowed between when the cook receives an order and when it must be ready?

_____

_____

13. Describe the production sheet used in this area. How are production needs forecast? **Attach a production sheet if available.**

_____

_____

_____

_____

14. What are the cleaning and sanitation duties of the entrée cook?

_____

_____

_____

_____

_____

15. Sketch out the area in which the entrée cook works. Indicate all equipment and personnel. Use arrows to show work flow.

16.

17.

Name: _____

Date(s): _____

Department/Location: _____

Source of Information: _____

1. Who prepares the vegetables and starches in your operation? What are the scheduled hours for this cook (or cooks)?

   _____

   _____

   _____

2. List the cooking equipment and processing equipment used in this area.

   _____

   _____

   _____

   _____

3. Describe the preparation of vegetables and starches that goes on in this area.

   _____

   _____

   _____

   _____

   _____

   _____

4. Name five popular vegetable dishes on the menu.

   _____

   _____

   _____

   _____

   _____

   _____

Name five popular starchy dishes on the menu.

_____

_____

_____

_____

_____

5. What principal methods are used to cook vegetables?

_____

_____

_____

_____

What principal methods are used to cook potatoes and other starchy dishes?

_____

_____

_____

_____

6. Are vegetables and starches cooked to order or in advance?

_____

_____

_____

If cooked ahead, how are they held until service?

_____

_____

_____

_____

7. Are any special spices or sauces used to accent vegetable or starchy dishes? If so, describe them.

_____

_____

_____

_____

8. How are vegetables and starchy dishes garnished?

_____

_____

_____

_____

9. Describe the production sheet used in this area. How are production needs forecast? **Attach a production sheet if available.**

_____

_____

_____

_____

10. What are the cleaning and sanitation duties of the vegetable cook?

_____

_____

_____

_____

_____

11. Sketch out the area in which the vegetable cook works. Indicate all equipment and personnel. Use arrows to show work flow.

12.

13.

Name: _____

Date(s): _____

Department/Location: _____

Source of Information: _____

1. Who does the cooking in the fast food section of the restaurant or the fast food restaurant?

_____

_____

What are the scheduled hours for this cook (or cooks)?

_____

_____

2. What is the most popular food?

_____

Is there a menu item that is unique to your operation? If so, describe it.

_____

3. List the menu items prepared by the cook.

_____

_____

_____

_____

_____

4. What menu items are prepared:

on the grill?

_____

_____

_____

_____

using the broiler?

_____

_____

_____

_____

_____

by microwaving?

_____

_____

_____

_____

_____

by deep-fat frying?

_____

_____

_____

_____

_____

5. List any convenience products used.

_____

_____

_____

6. List the cooking equipment used in this area.

_____

_____

_____

7. Are any fast foods made that are intended to appeal to customers looking for something nutritious? If so, describe them.

_____

_____

_____

_____

8. Describe the production sheet used in this area. How are production needs forecast? **Attach** a production sheet if available.

_____

_____

_____

_____

9. What are the cleaning and sanitation duties of the fast food cook?

_____

_____

_____

_____

_____

10. Sketch out the area in which the fast food is prepared. Indicate all equipment and personnel. Use arrows to show work flow.

11.

12.

Name: _____

Date(s): _____

Department/Location: _____

Source of Information: _____

1. Who prepares the baked goods in your operation?

   _____

   _____

   What are the scheduled hours for this cook (or cooks)?

   _____

   _____

   _____

2. What is the most popular baked good?

   _____

   Is there a baked item that is unique to your operation? If so, describe it.

   _____

3. List the baked goods made from scratch.

   _____

   _____

   _____

   _____

   _____

   _____

   List the baked goods made from mixes or frozen dough.

   _____

   _____

   _____

   _____

   _____

4. List the cooking equipment and tools commonly used in this area.

_____

_____

_____

_____

5. How are the baked goods garnished?

_____

_____

_____

_____

6. What size cake is baked?

_____

_____

How many portions do you get from one cake?

_____

7. What size pie is baked?

_____

_____

How many portions do you get from one pie?

_____

8. What are the portion sizes for other baked goods, such as muffins?

_____

_____

_____

_____

9. Are any baked goods made that are intended to appeal to customers looking for something nutritious? If so, describe them.

_____

_____

_____

_____

10. Describe the production sheet used in this area. How are production needs forecast? **Attach** a production sheet if available.

_____

_____

_____

_____

11. What are the cleaning and sanitation duties of the baker?

_____

_____

_____

_____

_____

12. Sketch out the area in which the baker works. Indicate all equipment and personnel. Use arrows to show work flow.

13.

14.

Name: _____

Date(s): _____

Department/Location: _____

Source of Information: _____

1. Who prepares salads, sandwiches, and other cold foods in your operation?

   _____

   _____

   What are the scheduled hours for this cook (or cooks)?

   _____

   _____

   _____

2. What is the most popular item produced in the pantry?

   _____

3. What categories of cold foods (salads, sandwiches, cold platters, cold appetizers, salad dressings, dips, desserts) are prepared in this area? Give three examples of each category.

   _____

   _____

   _____

   _____

   _____

   _____

4. Does the pantry make any trays or platters for buffets or catered events? If so, describe.

   _____

   _____

   _____

   _____

   _____

   _____

5. Give examples of four different garnishes used.

_____

_____

_____

_____

6. List the equipment used in this area.

_____

_____

_____

_____

7. Are any pantry items made that are intended to appeal to customers looking for something nutritious? If so, describe them.

_____

_____

_____

_____

8. Describe the production sheet used in this area. How are production needs forecast? **Attach a production sheet if available.**

_____

_____

_____

_____

9. What are the cleaning and sanitation duties of the personnel in the pantry?

_____

_____

_____

_____

_____

_____

10. Sketch out the area in which cold food preparation takes place. Indicate all equipment and personnel. Use arrows to show work flow.

11.

12.

1. Schedule reservations.

2. Check table settings for correctness and tray stations for adequacy of supplies before opening the dining room.

3. Check that all service personnel are present and properly dressed and groomed per policy.

4. Assign stations to all service personnel.

5. Inform service personnel of any menu specials, menu changes, or other information of which they need to be aware.

6. Greet guests upon arrival and escort them to their table, taking into account balanced seating and any guest requests.

7. Assist with seating of guests and provide menu, name of server, and any other appropriate information.

8. Oversee dining room service, assisting servers and dining room attendants as appropriate, inquiring of guests about their dining satisfaction, and resolving guest complaints.

9. Write and post the service personnel schedule in a designated location.

10. Verify and total hours worked/nonworked for payroll.

11. Complete, or assist in completing, performance appraisals for service personnel.

12. Train new service personnel.

13. Periodically run meetings of service personnel to review policies and procedures, discuss new policies and procedures, receive employee input, perform training, and so forth.

14. _____

15. _____

Name: _____

Date(s): _____

Department/Location: _____

Source of Information: _____

1. What functions does the dining room manager perform before the dining room opens? Does it include scheduling reservations? **Attach** a reservations form if available.

   _____
   _____
   _____
   _____
   _____

2. Does the assignment of stations to service personnel appear to be fair and equitable?

   _____
   _____
   _____
   _____

3. Is there a premeal meeting? If so, what happens at this meeting?

   _____
   _____
   _____
   _____

4. Once the dining room is open, are guests seated promptly and courteously?

   _____
   _____

5. What information does the dining room manager impart to the guests during seating?

   _____
   _____
   _____

6. List the activities the dining room manager performs during dining room service.

_____

_____

_____

_____

_____

7. Does the dining room manager frequently check on guest satisfaction by speaking directly to the guests?

_____

_____

8. How are guest complaints handled?

_____

_____

_____

_____

9. How far in advance is the schedule for service personnel posted, and where is the schedule posted?

_____

_____

_____

10. How many servers and dining room attendants are normally scheduled for the different meal periods (breakfast, lunch, dinner)?

_____

_____

_____

11. Does the dining room manager verify and total hours worked and nonworked for payroll? If so, describe the process used.

_____

_____

_____

_____

12. Does the dining room manager complete, or assist in completing, performance appraisals? If so, describe the process.

_____

_____

_____

_____

_____

13. Describe the extent to which the dining room manager trains new service personnel.

_____

_____

_____

_____

_____

14. Does the dining room manager conduct regular meetings (besides the premeal meetings) with the service personnel? If so, what occurs at these meetings?

_____

_____

_____

15.

16.

1. Perform opening duties such as setting tables and stocking the server station(s).

2. Serve guests and meet their needs to ensure a pleasant dining experience and a return visit.

   - Greet guests promptly and courteously.

   - Inform guests of specials and menu changes.

   - Make suggestions and answers guests' questions regarding food, beverages, and service.

   - Accurately describe all menu items and beverages available.

   - Take guests' order.

   - Relay guests' orders to the kitchen and the bartender as appropriate.

   - Serve food and beverages promptly and according to service policy.

   - Serve and refill guests' rolls, butter, other condiments, and water glasses as appropriate and as needed.

   - Remove used dishes, silverware, and glasses between courses and at the end of the meal.

   - Empty ashtrays or provide new ashtrays as necessary.

   - Check with guests for satisfaction with food and service.

   - Handle guest complaints.

   - Complete the guest check properly and handle cash and credit cards according to procedures.

3. Serve alcohol responsibly.

4. Perform side duties or sidework such as resetting tables, restocking server stations, or preparing beverages.

5. Perform closing duties.

6. Maintain cleanliness and sanitation of the dining room, including tables, chairs, floors, and windows.

7. Communicate and coordinate with managers, other servers, and dining room attendants to ensure guest satisfaction.

8. _____

9. _____

Name: _____

Date(s): _____

Department/Location: _____

Source of Information: _____

1. In the space below, draw one place setting as it is done in your operation. Draw the proper placement of all dishes, utensils, and glassware.

2. Draw the floor plan for the dining room, including stations, below.

3. How many persons is each server expected to serve?

_____

_____

4. Are the servers rotated or assigned stations, and on what basis is this accomplished?

_____

_____

_____

_____

5. How many types of menus are used in your dining room? During what time periods are they used?

_____

_____

_____

6. List the opening duties for servers.

_____

_____

_____

_____

_____

7. What items are stored or kept at the server's station?

_____

_____

_____

8. Describe what happens when the server first greets the guests at the table. What is the procedure?

_____

_____

_____

_____

_____

9. Does the server make suggestions to guests, such as wine with their meal or dessert selections?

_____

_____

_____

10. How are menu items written on the guest check? What abbreviations are used? Is there a list of standard abbreviations. **Attach** a list if available.

_____

_____

_____

_____

_____

_____

11. Does the server use a system to remember which person at the table ordered which item? If so, describe it.

_____

_____

_____

_____

12. What is the procedure for relaying the guests' order to the kitchen and the bartender?

_____

_____

_____

13. What is the procedure for picking up the food from the kitchen? Does the server have to do any food preparation or garnishing? If so, describe it.

_____

_____

_____

_____

_____

14. Is there a written sequence of service procedures available to the servers? Describe the sequence of procedures or **attach** a copy of the procedures.

_____

_____

_____

_____

_____

_____

_____

_____

_____

_____

15. Does the server ask the guests if everything is satisfactory? If so, when?

_____

_____

_____

16. How are guest complaints handled?

_____

_____

_____

_____

_____

17. What is the extent of empowerment given to servers? Describe it.

_____

_____

_____

_____

_____

_____

_____

_____

18. How do you complete a guest check at the end of the meal? **Attach** a guest check if available.

_____

_____

_____

_____

_____

_____

_____

19. How can guests pay for their meal? Describe the procedure for handling each form of payment.

_____

_____

_____

_____

_____

_____

20. Does each server keep his or her own tips, or are tips pooled? If there is any split on tips, with whom are they split, and what is the percentage of the split?

_____

_____

_____

21. What guidelines are used when serving alcohol?

_____

_____

_____

_____

_____

22. What sidework is performed during meal service?

_____

_____

_____

_____

_____

23. Describe the closing duties of the server.

_____

_____

_____

_____

_____

24. What procedures are followed when the following situations occur?

A large order is ready to be picked up from the kitchen, but one item is missing.

_____

_____

_____

An order is ready to be picked up from the kitchen, but it is not hot enough or attractively presented.

_____

_____

_____

A guest requests a food item that is not on the menu.

_____

_____

_____

A guest has had too much alcohol.

_____

_____

_____

Guests need to be seated, but there are no clean tables.

_____

_____

_____

A guest tries to pay with a suspected lost or stolen credit card.

_____

_____

_____

A guest becomes ill.

_____

_____

_____

25. Does each server know what amount of sales he or she generates? Is a certain sales volume expected of each server?

_____

_____

_____

_____

26. Are there any special food or beverage promotions? Are servers given training or offered incentives to sell these items?

_____

_____

_____

_____

27. What shifts do servers work, and what breaks are allowed?

_____

_____

_____

_____

28. Who makes out the servers' schedule, and how far in advance is it posted? Is the schedule permanent or rotating?

_____

_____

_____

29. What is the procedure and accounting system for issuing guest checks?

_____

_____

_____

30. Are there any control procedures for linens such as tablecloths and napkins? If so, describe them.

_____

_____

_____

31. What types of training are given to servers before starting work, and how extensive is the training program?

_____

_____

_____

_____

_____

32. What is the dress code for servers?

_____

_____

_____

33. If uniforms are used, who furnishes them and who is responsible for laundering them?

_____

_____

_____

34.

35.

Name: _____

Date(s): _____

Department/Location: _____

Source of Information: _____

1. With how many servers will one dining room attendant work?

   _____

   _____

2. What duties are the dining room attendants required to perform to restock the server's station or beverage station?

   _____

   _____

   _____

   _____

   _____

3. What are the dining room attendant's duties and responsibilities for serving and refilling guests' rolls, butter, other condiments, and water glasses?

   _____

   _____

   _____

   _____

4. What are the procedures for the dining room attendant for clearing and resetting tables during meal service and after guests leave?

   _____

   _____

   _____

   _____

   _____

5. When dirty dishes and so forth are taken to the dishwashing area, how are these items to be sorted?

_____

_____

_____

_____

6. How does the dining room attendant handle dirty ashtrays?

_____

_____

_____

7. Does the dining room attendant prepare beverages? If so, describe the procedures.

_____

_____

_____

_____

_____

8. Describe any side duties not already mentioned.

_____

_____

_____

_____

_____

9. Describe the closing duties.

_____

_____

_____

_____

10. What shifts do dining room attendants work, and what breaks are allowed?

_____

_____

_____

11. Who makes out the attendant's schedule, and how far in advance is it posted? Is the schedule permanent or rotating?

_____

_____

_____

12. Who is the supervisor for the dining room attendants?

_____

_____

13. Are the servers required to split tips with the dining room attendants as a set amount or as a percent of their total tips?

_____

_____

_____

14. What types of training are given to dining room attendants before starting work, and how extensive is the training program?

_____

_____

_____

_____

_____

15. What is the dress code for dining room attendants?

_____

_____

_____

16. If uniforms are used, who furnishes them and who is responsible for laundering them?

_____

_____

_____

17.

18.

1. Take guests' orders.

2. Fill guests' orders.

3. Collect cash, check, and/or charge payments from guests.
   - Make change correctly for cash transactions.
   - Follow procedures for verifying identification and processing of checks.
   - Follow procedures for processing credit card charges.

4. Respond to guest complaints.

5. _____

6. _____

Name: _____

Date(s): _____

Department/Location: _____

Source of Information: _____

1. Make a sketch of the counter service area.

2. Describe the steps involved in taking a guest's order.

_____

_____

_____

_____

_____

_____

_____

3. Which food orders need to be given to the cook (or cooks)?

_____

_____

_____

_____

_____

4. Where does the server have to go to get food orders filled? Give the name of each station and the foods there.

_____

_____

_____

_____

_____

_____

_____

_____

_____

5. How many sizes of cold drinks are available?

_____

_____

_____

How many sizes of hot drinks are available?

_____

_____

_____

6. What drinks are available from the drinks station?

_____

_____

_____

_____

7. Describe the procedures for collecting payment.

_____

_____

_____

_____

_____

_____

_____

_____

8. What other duties does the server perform?

_____

_____

_____

_____

_____

_____

_____

_____

_____

9. What shifts do servers work, and what breaks are allowed?

_____

_____

_____

10. Who makes out the servers' schedule, and how far in advance is it posted? Is the schedule permanent or rotating?

_____

_____

_____

11. What types of training are given to servers before starting work, and how extensive is the training program?

_____

_____

_____

_____

_____

_____

12. What is the dress code for servers?

_____

_____

_____

13. If uniforms are used, who furnishes them and who is responsible for laundering them?

_____

_____

_____

_____

14.

15.

1. Prepare stations for service.

2. Stock plates, cups, carryouts, and other dishes and disposables.

3. Practice the correct method of greeting guests, filling orders, and keeping guests moving at maximum speed.

4. Maintain portion control.

5. Fill salad bars and other stations.

6. Break down stations at the end of service.

7. Make sure food is stored properly before, during, and after service.

8. Garnish foods.

9. Perform sanitation duties and maintain sanitation standards.

10. _____

11. _____

Name: _____

Date(s): _____

Department/Location: _____

Source of Information: _____

1. Make a sketch below of the cafeteria and dining areas. Be sure to indicate the name of each cafeteria station such as the salad bar.

2. Who supervises/manages the cafeteria servers?

   _____

   _____

3. How is each station set up for service?

   _____

   _____

   _____

   _____

   _____

   _____

   _____

   _____

   _____

   _____

   _____

   _____

4. What is the portion size for sandwiches?

_____

_____

What is the portion size for hamburgers?

_____

_____

What is the portion size for hot vegetables?

_____

_____

5. What are the most popular menu items?

_____

_____

_____

_____

_____

6. How are foods garnished and kept looking fresh?

_____

_____

_____

_____

7. How many sizes of cold drinks are available?

_____

_____

_____

How many sizes of hot drinks are available?

_____

_____

_____

8. What drinks are available from the drinks station?

_____

_____

9. How is each station broken down at the end of service?

_____

_____

_____

_____

_____

_____

_____

_____

_____

_____

_____

10. How are guest complaints handled?

_____

_____

_____

_____

_____

11. What are the cafeteria hours?

_____

_____

_____

_____

12. What shifts do servers work, and what breaks are allowed?

_____

_____

_____

_____

13. Who makes out the servers' schedule, and how far in advance is it posted? Is the schedule permanent or rotating?

_____

_____

_____

14. What types of training are given to servers before starting work, and how extensive is the training program?

_____

_____

_____

_____

_____

15. What is the dress code for servers?

_____

_____

_____

16. If uniforms are used, who furnishes them and who is responsible for laundering them?

_____

_____

_____

17.

18.

1. Receive and verify the cash bank at the beginning of a shift.

2. Open the cash register according to procedures.

3. Obtain pricing information, such as menu specials and revised prices.

4. Ring up guest checks according to procedures.

5. Collect cash, check, and/or charge payments from guests.
   - Make change correctly for cash transactions.
   - Follow procedures for verifying identification and processing of checks.
   - Follow procedures for processing credit card charges.

6. Follow closing procedures, including counting money, checks, and charge vouchers.

7. Reconcile receipts with total sales.

8. _____

9. _____

Name: _____

Date(s): _____

Department/Location: _____

Source of Information: _____

1. How many cash registers does your operation have?

   _____

   Where are they located?

   _____

   _____

   _____

   _____

2. Who is responsible for opening/closing the cash registers/terminals and accepting guest payments? List the employee's names and titles.

   _____

   _____

   _____

   _____

   _____

   _____

3. Describe the procedure for opening a cash register.

   _____

   _____

   _____

   _____

   _____

   _____

   _____

   _____

   _____

4. Describe how guest checks are rung up.

_____

_____

_____

_____

_____

_____

5. What taxes are added to the guest check? What do these taxes cover, and how much is each one?

_____

_____

_____

_____

_____

6. Describe how the following payments are processed.

Cash: _____

_____

_____

_____

_____

Check: _____

_____

_____

_____

_____

Charge: _____

_____

_____

_____

_____

7. Describe the procedure for closing a cash register.

_____
_____
_____
_____
_____
_____
_____
_____

8. Who reconciles the receipts with the total sales rung up?

_____
_____
_____

9. What specific procedures are used in your operation to prevent employees from stealing money from the register?

_____
_____
_____
_____
_____
_____

10.

11.

1. Set up a bar.

2. Do a physical bar inventory.

3. Requisition liquor and supplies for the bar.

4. Specify where bar brands, call brands, and liqueurs are kept.

5. Mix and serve alcoholic and nonalcoholic drinks according to house procedures.

6. Garnish alcoholic and nonalcoholic drinks.

7. Show and tell about special house drinks.

8. List the size and type of glasses to use.

9. Wash all glasses with the proper sink setup.

10. Ring up beverage sales.

11. Inventory liquor storage area and write up purchase orders.

12. Receive and issue liquor.

13. Trace the liquor control system from purchase to cash turn-in.

14. Calculate daily bar cost percentage.

15. Calculate beverage prices.

16. Design a bar sales promotion.

17. Make an entertainment cost analysis.

18. Purchase and sell alcoholic beverages following all pertinent regulations.

19. _____

20. _____

Name: _____

Date(s): _____

Department/Location: _____

Source of Information: _____

1. Sketch below the room in which the front bar (or bars) is located. If the front bar has a pickup station, label the section of the bar used for that purpose. Include equipment that makes up the underbar, back bar, and pouring station.

2. Does your operation have a service bar? If so, where is it located?

   _____
   _____
   _____
   _____

3. Is there entertainment in the bar? If so, what type of entertainment and how many nights per week?

   _____
   _____
   _____
   _____
   _____

4. Describe the clientele that frequents the bar.

_____

_____

_____

_____

_____

5. Besides the bartender, who takes beverage orders and serves beverages?

_____

_____

_____

_____

If cocktail servers are used, how many customers is each responsible for serving?

_____

6. Describe the system used to account for and ring up customers' drinks. Is each drink paid for when served, or is a tab kept?

_____

_____

_____

_____

_____

_____

7. List the primary functions the bartender (or bartenders) performs.

_____

_____

_____

_____

_____

_____

_____

_____

_____

8. Who do the bartender and beverage servers report to? Give name and title.

_____

_____

_____

9. What personal qualities are essential for a bartender?

_____

_____

_____

_____

_____

_____

10. How often is an inventory taken at the bar and in the storeroom?

_____

_____

_____

Is a requisition system used to bring the bar inventory up to par? If not, describe the system used.

_____

_____

_____

_____

11. What are the 5 most popular or frequently poured drinks the bartender serves? What is the standard drink size and price for each?

_____

_____

_____

_____

_____

Is any one of the drinks above considered to be a specialty of the house?

_____

_____

12. How are drink prices determined and who is responsible for setting prices?

_____

_____

_____

_____

_____

_____

13. Are standardized recipes used? To what extent? **Attach** a sample recipe.

_____

_____

_____

What type of measuring device is used for drinks?

_____

_____

_____

14. Are automatic dispensing systems used at the bar? If so, for which beverages?

_____

_____

_____

_____

_____

_____

_____

15. What are the advantages and disadvantages of using automatic liquor-dispensing systems?

_____

_____

_____

_____

_____

_____

_____

16. What brands of liquors are kept in the speed rail?

_____

_____

_____

_____

17. Give five examples of garnishes used on drinks.

_____

_____

_____

_____

18. List the types of tumblers, footed glasses, and stemware used and for which drinks each glass is used.

_____

_____

_____

_____

_____

_____

_____

_____

19. Describe the procedure and equipment for washing glasses.

_____

_____

_____

_____

20. What measures are taken during service to ensure adequate sanitation?

_____

_____

_____

_____

_____

21. What controls are set up to prevent employee theft of liquor and cash?

_____

_____

_____

_____

_____

_____

_____

_____

22. Who is responsible for determining how much to order and placing and receiving the beverage orders?

_____

_____

_____

_____

_____

23. What is the typical bar cost percentage (the ratio of cost of liquor used to the total dollar sales)?

_____

_____

_____

_____

24. Describe the marketing and promotion activities in this area.

_____

_____

_____

_____

_____

25.

26.

Name: _____

Date(s): _____

Department/Location: _____

Source of Information: _____

1. What licenses and/or permits does the establishment have that allows it to sell alcoholic beverages?

   _____

   _____

   _____

   _____

   _____

   From where were these licenses and permits issued, and how much did each cost?

   _____

   _____

   _____

   _____

   _____

   How often do licenses and permits have to be renewed and at what cost?

   _____

   _____

   _____

   _____

   _____

2. What types of alcoholic beverages can be sold?

   _____

   _____

   _____

   _____

   _____

3. When can you sell alcoholic beverages?

4. To whom can you sell alcoholic beverages?

5. Is the state you are in considered a control state (monopoly state) or a license state?

6. Discuss all regulations that affect where you purchase alcoholic beverages.

7. Discuss all regulations that affect how you purchase alcoholic beverages.

_____

_____

_____

_____

_____

_____

_____

_____

_____

_____

_____

8. What measures are taken to prevent guests from driving home drunk?

_____

_____

_____

_____

_____

_____

_____

_____

9.

10.

1. Work the dining room as a wine steward.

2. Demonstrate correct wine-serving procedure.

3. Make wine suggestions.

4. Explain the information shown on domestic and imported labels.

5. Complete a wine inventory.

6. Inventory the wine storage area and write up purchase orders.

7. Receive and issue wine.

8. Determine wine prices.

9. _____

10. _____

Name: _____

Date(s): _____

Department/Location: _____

Source of Information: _____

1. List the five most popular wines served and their prices.

   _____

   _____

   _____

   _____

   _____

2. Do your guests prefer American or European wines?

   _____

   _____

   Is anything done to promote sales of one over the other? If yes, what and why?

   _____

   _____

   _____

   _____

   _____

   How do you account for the popularity of one over the other?

   _____

   _____

   _____

3. Is a separate wine list available to guests, or is the list of wines included on the menu? **Attach** a list of wines if available.

   _____

   _____

   _____

If a wine list is available, in what way are the wines listed?

_____

_____

_____

4. Define the following wine terms, some of which appear on wine labels.

Still wine: _____

_____

Sparkling wine: _____

_____

Champagne: Brut, Extra Sec, Sec, Demi-Sec, Doux: _____

_____

Table wine: _____

_____

Fortified wine: _____

_____

Aperitif: _____

_____

Dry: _____

_____

Vintage date: _____

_____

Appellation Contrôlée: _____

_____

Denominazione di Origini Controllata (DOC): _____

_____

Qualitätswein: _____

_____

Mis en bouteille au château: _____

_____

5. How are wines promoted to guests?

_____

_____

_____

_____

_____

_____

6. Is there an employee, such as a wine steward, responsible for selling and serving wine? If so, describe his or her job duties.

_____

_____

_____

_____

_____

Does this employee wear a special uniform or use any special equipment? If so, describe it.

_____

_____

_____

_____

If there is no wine steward, have the servers been trained on the selling and service of wines? If so, describe the training.

_____

_____

_____

_____

_____

7. Does the wine steward and/or servers make a commission for wine sold? If so, how is the amount of the commission determined?

_____

_____

_____

_____

8. Explain the procedure for serving wine, including the types of glasses used and any equipment used.

_____

_____

_____

_____

_____

_____

9. Give an estimate of the dollar volume for the amount of wine sold per week.

_____

_____

What nights are best for wine sales?

_____

_____

About what percentage of the guests order wine?

_____

_____

10. What is the procedure for purchasing wine?

_____

_____

_____

_____

_____

Who is responsible for purchasing wine? Give name and title.

_____

_____

11. Where is wine stored?

_____

_____

_____

_____

How often is it inventoried?

_____

_____

12. Are bin cards used in the wine storage area?

_____

_____

If yes, explain how they are used.

_____

_____

_____

_____

_____

13. How is wine issued from the storage area?

_____

_____

_____

_____

_____

14. What volume of wine, by bottle count and/or dollar volume, is kept on hand?

_____

_____

_____

_____

_____

15. What are the various sizes of containers in which wine is purchased?

_____

_____

_____

_____

_____

Which size is most popular?

_____

_____

_____

16. Discuss with your manager the advantages and disadvantages of glass, bulk, and bottle wine sales.

_____

_____

_____

_____

_____

_____

_____

_____

17. Who in the operation establishes the sales price for wine?

_____

_____

What is the basis for establishing the wine price? Give the approximate markup.

_____

_____

_____

_____

18.

19.

1. Handle catering requests from initial contact by groups to breaking down the catering event and storing all equipment and supplies.

2. Determine food cost and labor cost on catered events.

3. Set catering prices.

4. Calculate the profitability of catered events.

5. Serve at buffets and banquets.

6. _____

7. _____

Name: _____

Date(s): _____

Department/Location: _____

Source of Information: _____

1. Do you have standard menus written for catered events? Are menus often customized to the group? Describe the menus used for catering buffets, banquets, and other catered events. **Attach** sample menus if available.

_____

_____

_____

_____

_____

_____

_____

_____

_____

_____

_____

Are theme menus available? If so, describe them.

_____

_____

_____

_____

_____

Who writes the menus? Give name and title.

_____

_____

_____

_____

_____

2. How many catering rooms or spaces are available for parties? What is the capacity of each room for the various types of functions that may be held?

_____

_____

_____

_____

_____

_____

_____

3. What is the largest group you can serve?

_____

_____

What is the smallest group size you require for catering?

_____

_____

4. List the employees (include job titles) who are involved in catering activities, and describe the functions each person performs. If you are working in a catering department, give the organizational structure.

_____

_____

_____

_____

_____

_____

_____

_____

_____

_____

5. Who does the room and table setup for catered events? Describe any special forms used to help in room and table setup.

_____

_____

_____

_____

_____

_____

_____

_____

6. For buffet service, sketch how the buffet tables are decorated and the types of centerpieces used.

7. What are the advantages of buffet service for the operation and for the guests?

_____

_____

_____

_____

_____

_____

_____

8. What functions do serving personnel perform during buffets?

_____

_____

_____

_____

_____

9. Who prepares the food for catered events? Do any of the regular cooking staff prepare these foods?

_____

_____

_____

_____

_____

10. What is the normal food cost on a catered event?

_____

_____

What is the normal labor cost on a catered event?

_____

_____

How are catering prices set? How do they compare with regular menu prices?

_____

_____

_____

_____

_____

_____

How profitable overall are catered events?

_____

_____

11. How much monthly and annual dollar volume is done in catering?

_____

_____

12. What arrangements are made for payment? Does the operation require a deposit? If yes, how much and when is it paid? When is full payment due? Are credit cards accepted?

_____

_____

_____

_____

13. When must the guarantee of the number to attend be given?

_____

_____

What deviation is allowed on this number?

_____

_____

_____

14. Who schedules catering personnel?

_____

_____

When is their schedule posted?

_____

_____

_____

15.

16.

Training Guide—Banquet Service

Name: _____

Date(s): _____

Department/Location: _____

Source of Information: _____

1. Where do you obtain servers for banquets?

_____

_____

_____

_____

_____

2. How are banquet servers paid? Do banquet servers share the tip? If so, how much of a tip is automatically added to the banquet bill? How is the tip divided among the workers?

_____

_____

_____

_____

3. Do the banquet servers set the tables? How is the work divided? How much time is allowed?

_____

_____

_____

_____

4. What other setup work is done by the banquet servers? What instructions are they given? **Attach** sample instructions if available.

_____

_____

_____

_____

What breakdown and cleanup work is done by the banquet servers? What instructions are given?

_____

_____

_____

_____

5. Does the operation use any special dishes or silver or a different pattern of china or silver for banquets? If yes, how is it specially handled to keep it separate from the regular stock?

_____

_____

_____

6. How many covers is a server expected to serve during a banquet?

_____

_____

Is there assistance for servers, during service such as from runners, bussers, or beverage servers?

_____

_____

7. Is there a special menu for banquets? **Attach** a sample menu if available.

_____

_____

_____

8. How is the food plated and transported to the banquet room?

_____

_____

_____

_____

9. What system is used to keep count of the total number of people present and the total number of plates served?

_____

_____

_____

_____

10. How much and what type of training is given to new servers on the banquet crew?

_____

_____

_____

_____

_____

11. What is the uniform for banquet servers? Is it different from the dining room servers uniform?

_____

_____

_____

_____

12. When is the banquet crew fed? Are they fed from the banquet menu or other menus?

_____

_____

_____

13.

14.

1. Load all types of machines properly.

2. Do cash count and turn-in.

3. Determine the number and type of machines needed for various locations.

4. Contact appropriate personnel for repairs.

5. Evaluate vending contracts.

6. Oversee/enforce vending contracts.

7. Calculate sales and profit requirements for machines.

8. _____

9. _____

Name: _____

Date(s): _____

Department/Location: _____

Source of Information: _____

1. How many of each of the following vending machines are available for use?

   Hot drink machines   _____

   Cold drink machines   _____

   Snack machines   _____

   Food/meal machines   _____

   Milk machines   _____

   Ice cream machines   _____

   Other   _____

2. What is the product capacity of each type of machine?

   Hot drink machines   _____

   Cold drink machines   _____

   Snack machines   _____

   Food/meal machines   _____

   Milk machines   _____

   Ice cream machines   _____

3. How many persons is each machine considered capable of serving?

   Hot drink machines   _____

   Cold drink machines   _____

   Snack machines   _____

   Food/meal machines   _____

   Milk machines   _____

   Ice cream machines   _____

4. Do the soft drink machines offer canned or postmix soft drinks?

_____

_____

Which one is more profitable?

_____

_____

5. Are brand names used frequently in the machines? Why or why not?

_____

_____

_____

_____

_____

6. In which areas are vending machines located?

_____

_____

_____

_____

7. Who purchases foods and beverages from the vending machines?

_____

_____

_____

8. Are change machines and microwaves also available for customers?

_____

_____

What are the advantages and disadvantages of having change machines and microwaves available?

_____

_____

_____

_____

_____

9. What can a customer do if he or she loses money in a machine?

_____

_____

_____

_____

10. Who actually owns the vending machines?

_____

_____

11. Who refills the machines, collects the money, and maintains the vending machines?

_____

_____

_____

12. If a contractor operates the machines for you, describe the principal parts of your contract, including services the contractor must provide, how you are paid, and how much you are paid.

_____

_____

_____

_____

_____

_____

_____

_____

_____

13. What are the average weekly vending sales?

_____

_____

14. What is the schedule for refilling the machines and collecting the money?

_____

_____

_____

Explain briefly the checkout and inventory system used by employees who refill the machines.

_____

_____

_____

_____

_____

How much of vending products are disposed of on an average day because they are too old to be served?

_____

_____

_____

How does the employee who is refilling the machines know the product is too old?

_____

_____

_____

_____

How is the money amount verified?

_____

_____

_____

_____

How many hours per week are required to refill and maintain the machines and collect money?

_____

_____

15. How much do the various types of machines cost?

Hot drink machines   _____

Cold drink machines   _____

Snack machines   _____

Food/meal machines   _____

Milk machines   _____

Ice cream machines   _____

16. What is the average life of a vending machine?

_____

_____

17. What income must a machine have to be profitable?

_____

_____

_____

Which types of machines are most profitable?

_____

_____

_____

18. To what extent is breakage and vandalism of machines a problem?

_____

_____

_____

What is done to protect the machines, the money, and the product?

_____

_____

_____

_____

19. What are the most frequent malfunctions in vending machines?

_____

_____

_____

_____

Who contacts the company in charge of repairs?

_____

20.

21.

1. List the location of all cleaning supplies.

2. Use the dishmachine properly: prepare it for operation, scrape dishes, sort dishes, rack dishes or put them on the conveyor belt, operate the machine at proper temperatures, replenish the soap dispenser and rinse additive when needed, remove clean dishes and store them properly, break down the machine and clean it.

3. Clean pots, pans, and smallwares using a three-compartment sink, ensuring that the sanitizer is at the appropriate level.

4. Keep dishwashing and potwashing areas clean, including storage areas, all equipment, walls, and floors.

5. Remove trash appropriately from the kitchen.

6. Use sanitary procedures to clean garbage cans, floors, walls, and equipment throughout the kitchen.

7. Clean and polish all lowerators and equipment used to store china, glassware, silverware, and like items.

8. Conduct a sanitation self-inspection.

9. Follow safety rules.

10. Follow appropriate procedures in case of fire or injury.

11. Conduct a safety self-inspection.

12. Maintain safety records.

13. Act to decrease the number of accidents.

14. _____

15. _____

Name: _____

Date(s): _____

Department/Location: _____

Source of Information: _____

1. Do employees working on sanitation jobs have their own supervisor? If so, please identify by name and title.

   _____

   Who schedules the employees who work in this area?

   _____

   When and where is their schedule posted?

   _____

   _____

   _____

2. Who is responsible for doing general kitchen cleaning, such as mopping floors and cleaning equipment?

   _____

   _____

   _____

   _____

   _____

   What shifts does this person (or persons) work?

   _____

   _____

   _____

   _____

3. Does this person have a daily cleaning schedule that indicates when to mop floors, take out the garbage, and so on? **Attach** the schedule if available.

   _____

   _____

Describe the work this person does.

_____

_____

_____

_____

_____

_____

4. Where are mops, cleaning supplies, and garbage bags stored?

_____

_____

_____

5. Describe the procedure for mopping floors, including names of all chemicals used and how they are mixed.

_____

_____

_____

_____

_____

_____

_____

6. Where is garbage removed to for pickup?

_____

_____

Who hauls the trash away, how often, and at what charge?

_____

_____

_____

7. Where are garbage cans cleaned?

_____

_____

How often are garbage cans cleaned?

_____

_____

What is the procedure for cleaning garbage cans?

_____

_____

_____

_____

_____

8. Is there a master cleaning schedule for the kitchen that indicates how often different areas and pieces of equipment are cleaned? **Attach** it if available.

_____

_____

How is this schedule implemented?

_____

_____

_____

9. How is the exhaust hood over the cooking equipment kept clean?

_____

_____

_____

How often is it cleaned?

_____

10. Does the operation do sanitation self-inspections? If so, describe how the inspections are done. **Attach** a self-inspection form if available.

_____

_____

_____

_____

_____

_____

11. Which regulatory agencies come in to do sanitation inspections? How often do they come in?

_____

_____

_____

_____

Did the operation pass the last inspection? What was the grade given?

_____

_____

12.

13.

Name: _____

Date(s): _____

Department/Location: _____

Source of Information: _____

1. How many warewashers are needed to staff the dishroom at the different operating hours? Note the times of each shift.

    _____

    _____

    _____

    _____

2. How many compartments does the dishmachine have?

    _____

3. Describe how to set up the dishmachine for operation.

    _____

    _____

    _____

    _____

    _____

    _____

    Where are the supplies for operating the dishmachine kept?

    _____

    _____

4. Describe the procedure of dishwashing from the scraping and sorting of dirty dishes to the storing of clean dishes.

    _____

    _____

    _____

    _____

5. How many temperature gauges are on the dishmachine?

_____

What does each gauge indicate?

_____

_____

How often are the gauges checked for accuracy?

_____

_____

Is a written temperature log kept? **Attach** *a sample if available.*

_____

6. Where is the soap dispenser located on the dishmachine?

_____

How often is the soap dispenser checked?

_____

_____

Describe any other dispensers on the dishmachine.

_____

_____

_____

7. Describe the procedure for breaking down and cleaning the dishmachine.

_____

_____

_____

_____

_____

_____

8. Does the operation use a water-softening machine to soften hard water?

_____

_____

If not, do the dishwashers regularly delime the machine? How is this done?

_____

_____

_____

_____

_____

9. Describe other duties performed by the dishwashers.

_____

_____

_____

_____

_____

_____

_____

10. Sketch out the area in which dishwashing takes place. Indicate all equipment and personnel. Use arrows to show work flow.

11.

12.

Name: _____

Date(s): _____

Department/Location: _____

Source of Information: _____

1. Is there one or more person whose job is primarily potwashing, or is potwashing performed by others such as the dishwasher or utility person?

   _____

   _____

   _____

2. What is the procedure for setting up the three-compartment sink (be sure to include the type of sanitizer used)?

   _____

   _____

   _____

   _____

   _____

   _____

   Where are the supplies for the sink stored?

   _____

   _____

3. How do you test that the sanitizing solution is adequate?

   _____

   _____

   _____

   _____

   _____

   How often is this done?

   _____

   _____

4. How often are the sinks drained and then set up again?

_____

_____

_____

_____

5. Where are the clean pots and pans stored?

_____

_____

_____

_____

_____

6. Describe other duties performed by the potwasher.

_____

_____

_____

_____

_____

_____

7. Sketch out the area in which dishwashing takes place. Indicate all equipment and personnel. Use arrows to show work flow.

8.

9.

Name: _____

Date(s): _____

Department/Location: _____

Source of Information: _____

1. Describe briefly the operation's safety policies and procedures. **Attach** the policies if available.

_____

_____

_____

_____

_____

_____

2. In case of fire, where are fire extinguishers located?

_____

_____

_____

_____

What type of fire extinguisher system is used in the exhaust system? Where do you turn it on?

_____

_____

_____

What is the procedure to follow in case of fire?

_____

_____

_____

_____

What regulations concerning fire ordinances and safety must the foodservice conform to?

_____

3. Are "Wet Floor" signs used to prevent falls?

_____

4. Is a first-aid kit available? Where is it located?

_____

_____

5. Are any staff trained to administer the Heimlich maneuver? List them and their positions.

_____

_____

Are any staff trained to give cardiopulmonary resuscitation (CPR)? List them and their positions.

_____

_____

6. In case an employee or guest becomes injured, what is the policy and procedure?

_____

_____

_____

_____

_____

Where are emergency phone numbers listed?

_____

_____

_____

7. Do employees lift properly, using their leg muscles (not their back muscles)?

_____

_____

8. What measures are taken to ensure that employees mix and use cleaning chemicals without harm?

_____

_____

_____

9. Describe the extent of formal training given employees regarding safety.

_____

_____

_____

_____

_____

_____

_____

10. Does the operation have a Safety Committee that meets regularly? If so, describe it.

_____

_____

_____

11. Describe the recordkeeping, if any, that is done when an accident occurs. **Attach** a sample form if available.

_____

_____

_____

_____

12. What are the most common types of accidents that occur?

_____

_____

_____

13. Is there a specific area or department where accidents are more likely to occur?

_____

_____

_____

_____

14. When was the last serious, or lost time, accident in your operation? Describe it.

_____

_____

_____

_____

_____

15. What is being done to reduce the number of accidents?

_____

_____

_____

_____

_____

_____

16. Are safety inspections held? If so, describe them, including who does the inspections.

_____

_____

_____

_____

_____

17.

18.

1. Identify security measures taken to protect employees.

2. Identify security measures taken to protect guests.

3. Identify security measures taken to prevent thefts by employees.

4. Identify security measures taken to prevent theft by guests.

5. Take appropriate actions in the event that guests walk out without paying and so on.

6. Report security incidents appropriately.

7. Help in security investigations.

8. _____

9. _____

Name: _____

Date(s): _____

Department/Location: _____

Source of Information: _____

1. Sketch below the organizational structure of the department in charge of security (if one exists).

2. Make a list of the major duties of each job in security.

_____

_____

_____

_____

_____

_____

_____

_____

_____

_____

3. What does security do to supervise entry and exit of employees?

_____

_____

Do employees have to enter and exit the restaurant or foodservice through designated doors? If so, explain why.

_____

_____

_____

_____

Do employees wear identification badges? If so, describe how the badges look and how they are used.

_____

_____

_____

_____

Are employees' packages and bags checked during entry and exit from the building? If so, why?

_____

_____

_____

Are there any situations in which an employee may leave with something that belongs to the restaurant or foodservice? If so, explain.

_____

_____

_____

_____

_____

4. Describe the process by which security investigates reported incidents such as theft of a guest's wallet.

_____

_____

_____

_____

_____

5. Discuss how incidents are reported. Do forms have to be filled out? If so, describe or **attach** the forms.

_____

_____

_____

_____

_____

_____

6. What is the most frequent incident that security becomes involved in?

_____

_____

7. Which incident has the greatest negative impact in terms of costs and lowered productivity?

_____

_____

_____

8. In cases of theft, which items are most often taken?

_____

_____

_____

9. Does security use any sensors, alarm systems, or special surveillance equipment to check for internal and external theft? If so, describe them.

_____

_____

_____

_____

_____

_____

_____

10. Are all employees who handle cash, checks, and credit cards trained to recognize forgery and counterfeit money, money orders, travelers' checks, and drivers' licenses, and how to deal with quick-change artists?

_____

_____

_____

_____

_____

11. List 10 other measures taken to prevent cash theft by robbers or by employees.

_____

_____

_____

_____

_____

_____

_____

_____

_____

_____

_____

12. Give three reasons that employees steal.

_____

_____

_____

_____

_____

13. If an employee sees a guest taking an item that belongs to the establishment, what is the employee supposed to do?

_____

_____

_____

_____

14. If an employee sees a guest leaving without paying, what is the employee supposed to do?

_____

_____

_____

_____

_____

15. Describe the areas in which security actively trains employees on security procedures.

_____

_____

_____

_____

_____

_____

_____

_____

16. Is security involved in any aspect of hiring any employee or employees for certain jobs? If so, describe its involvement.

_____

_____

_____

_____

_____

_____

17. List the areas for which the restaurant or foodservice has emergency plans. **Attach** a sample emergency plan if available.

_____

_____

_____

_____

_____

Describe what should happen in the event of a fire.

_____

_____

_____

_____

_____

_____

_____

_____

18. What type of public liability insurance is carried by the operation?

_____

_____

_____

How much coverage is provided?

_____

_____

_____

_____

What are the premiums?

_____

_____

19.

20.

1. Explain how accounting and finance are organized and the major duties of each job.

2. Interpret income statements.

3. Calculate the average check and average food cost per guest, average income per guest, and other averages as needed.

4. Calculate profitability, operating, and other ratios as needed.

5. Calculate standard food or beverage costs.

6. Identify internal controls.

7. Explain how operating budgets are prepared.

8. _____

9. _____

Name: _____

Date(s): _____

Department/Location: _____

Source of Information: _____

1. Does your company use an outside accounting firm for bookkeeping and accounting services?

   _____

   If so, what services does the firm provide and at what cost?

   _____

   _____

   _____

   _____

   _____

   _____

2. If an outside firm is not used, sketch below the organizational structure of the area in charge of accounting and finance.

3. Make a list of the major duties of each job in accounting and finance.

   _____

   _____

   _____

_____

_____

_____

_____

_____

_____

_____

_____

4. Who prepares income statements (or profit and loss statements) for the restaurant or foodservice? How often are they prepared?

_____

_____

_____

What important information does the income statement provide to the company?

_____

_____

_____

_____

_____

_____

_____

_____

_____

5. List the ratios the company generates and looks at, such as operating and profitability ratios.

_____

_____

_____

_____

_____

_____

_____

_____

_____

6. List the averages the company generates and looks at, such as the average check.

_____

_____

_____

_____

_____

_____

_____

_____

7. Explain how the company calculates a standard food or beverage cost.

_____

_____

_____

_____

_____

_____

_____

_____

8. Identify 10 internal controls.

_____

_____

_____

_____

_____

_____

_____

_____

_____

_____

_____

9. Describe the major steps in putting together an operating budget for the restaurant or foodservice.

_____

_____

_____

_____

_____

_____

_____

_____

_____

10. What period does the budget cover?

_____

11. Is the operating budget fixed or flexible?

_____

_____

_____

12. What categories do restaurant or foodservice managers budget for?

_____

_____

_____

_____

_____

_____

_____

_____

_____

_____

_____

13. Once an operating budget is determined, how is it used?

_____

_____

_____

_____

_____

_____

_____

14. What are the key elements in a capital budget?

_____

_____

_____

_____

_____

15.

16.

Introductory Worksheets for Institutional Foodservices

This chapter contains Introductory Worksheets for the following five segments of institutional foodservice.

- Business and industry foodservice
- School foodservice (includes schools from kindergarten to grade 12)
- Healthcare foodservice (includes hospitals and nursing facilities)
- College and university foodservice
- Retirement community foodservice

Institutional foodservices share significant similarities and differences with restaurants. The Training Goals and Training Guides for restaurants and foodservices (Chapter 3) will help you to understand operations. These worksheets will help you understand the uniqueness of these institutional settings. From feeding children school lunches to offering table service to retirees, institutional foodservice is rich in its diversity.

Name: _____

Date(s): _____

Department/Location: _____

Source of Information: _____

1. How many employees work at this location (don't include foodservice employees)?

   _____

   _____

2. Describe the types of employees at this location (such as factory workers, managers, researchers, and so on).

   _____

   _____

   _____

   _____

3. What dining services are offered, such as cafeteria, restaurant, catering, vending, and so on?

   _____

   _____

   _____

   _____

   _____

4. Which dining service is the most heavily used?

   _____

   _____

5. What is the average daily number of transactions for each dining service area?

   _____

   _____

   _____

   _____

   _____

6. What is the participation rate?

_____

_____

7. What is the average check for each dining service area?

_____

_____

_____

_____

8. Does the employer subsidize employee meals? If yes, state the percent subsidized.

_____

_____

_____

9. Which menu items are the most popular?

_____

_____

_____

10. Does the foodservice use many brand-name foods and beverages? Describe them.

_____

_____

_____

Are the brands used national brands and/or in-house brands?

_____

_____

_____

_____

11. Are special promotions or themed events planned? If so, give two examples.

_____

_____

_____

_____

_____

12. Ask the director or assistant director for his or her thoughts on three current concerns and issues facing business and industry foodservice.

_____

_____

_____

_____

_____

_____

_____

_____

_____

_____

13.

14.

Name: _____

Date(s): _____

Department/Location: _____

Source of Information: _____

1. How many students are in this school?

_____

_____

What are the ages and any outstanding characteristics of these students?

_____

_____

_____

_____

2. How many lunch meals are served daily?

_____

_____

3. Does the school foodservice participate in the National School Lunch Program?

_____

If yes, how many free meals, reduced-price meals, and full-price meals are served daily and at what price?

Free meals _____

Reduced-price meals _____

Full-price meals _____

To maintain compliance with this program, does the school follow the set meal pattern or do a nutritional analysis of meals to meet nutritional guidelines?

_____

_____

_____

4. Explain the basic elements of the National School Lunch Program. Include a brief discussion of the required government forms. **Attach** any forms if possible.

_____

_____

_____

_____

_____

_____

_____

_____

_____

_____

5. Does the school foodservice participate in the National School Breakfast Program?

_____

If yes, how many breakfasts are provided?

_____

6. If a student does not purchase lunch at school or receive a free lunch, what are his meal options?

_____

_____

_____

_____

Are vending machines available? What do they sell?

_____

_____

_____

_____

7. Which menu items are most popular with students?

_____

_____

_____

_____

Which menu items are least popular with students?

_____

_____

_____

8. Describe any nutrition education activities designed for the students. **Attach** any nutrition education materials if possible.

_____

_____

_____

_____

_____

_____

_____

9. In what manner does the school board make decisions or dictate policies and procedures for the school foodservice?

_____

_____

_____

_____

_____

_____

_____

_____

10. Besides the federal government, who reimburses the school for meals and in what amounts?

State _____

County _____

City/town _____

11. What is the food cost and labor cost of a lunch meal?

Food cost _____

Labor cost _____

12. How many employees are needed to produce:

100 lunches _____

200 lunches _____

300 lunches _____

13. Ask the director or assistant director for his or her thoughts on three current concerns and issues facing school foodservice.

_____

_____

_____

_____

_____

14.

15.

Name: _____

Date(s): _____

Department/Location: _____

Source of Information: _____

1. Describe the type of healthcare facility you are in. Is it a hospital, nursing home, or other type of operation?

   _____

   _____

   _____

2. How many beds are there?

   _____

   _____

   What is the average census?

   _____

   _____

3. What are the characteristics of the patient population, such as elderly or mainly poor?

   _____

   _____

   _____

   _____

4. Who inspects the foodservice operation and how frequently?

   _____

   _____

   _____

   _____

   _____

5. If you are in a nursing home, how has the Omnibus Budget Reconciliation Act of 1987 (commonly called OBRA) changed foodservice operations?

_____

_____

_____

_____

_____

_____

Patient Foodservice

6. Describe the key elements of how patient meals are assembled and delivered.

_____

_____

_____

_____

_____

_____

_____

_____

_____

7. Is tray service centralized or decentralized?

_____

_____

Sketch below the main tray assembly line, and indicate what each station does along the line.

8. Describe the type of equipment used to keep patient meals hot during delivery to their rooms.

_____

_____

_____

_____

_____

_____

9. How much time is permitted from when a tray is assembled to when it is delivered?

_____

_____

_____

Are most trays delivered within this standard delivery time?

_____

_____

_____

What are the most frequent problems that cause trays to be late?

_____

_____

_____

10. Is the patient menu selective or nonselective? **Attach** a patient menu if possible.

_____

_____

_____

How long is the menu cycle?

_____

_____

_____

List the menu pattern or menu categories for each meal, and give the typical number of choices in each menu category.

Breakfast: _____

_____

_____

_____

_____

_____

Lunch: _____

_____

_____

_____

_____

_____

_____

_____

Dinner: _____

_____

_____

_____

_____

_____

_____

_____

11. If a selective menu is used, how far in advance must the patient make his or her choices?

_____

_____

Who picks up the patient menus from the rooms?

_____

_____

12. How does the department forecast daily production needs?

_____

_____

_____

_____

_____

_____

13. How many patient meals are served daily?

_____

_____

14. Of the patient meals, how many are for regular diets?

_____

_____

How many are for modified or special diets?

_____

_____

15. List the types of modified or special diets available.

_____

_____

_____

_____

_____

_____

_____

_____

_____

_____

16. Are some of the patients served in a dining room? If so, describe how this type of service takes place.

_____

_____

_____

_____

_____

_____

_____

_____

_____

17. Does the department stock floor pantries so that nursing always has juice, crackers, and other items available for patients? If yes, list the foods stocked on the floors and state how often the pantries are restocked.

_____

_____

_____

_____

_____

_____

_____

_____

_____

_____

18. What contact is made with the patients by foodservice personnel to check the patient's satisfaction with the food? Who makes the contact? How often is the contact made? **Attach** any patient satisfaction forms if possible.

_____

_____

_____

_____

_____

_____

_____

_____

19. What is the average cost per patient meal?

Food cost: _____

Labor cost: _____

Nonpatient Foodservice

20. Is a cafeteria, or other style of foodservice, operated to provide food to health care employees and possibly visitors?

_____

_____

21. Make a sketch below of the cafeteria and dining areas. Be sure to indicate the name of each cafeteria station, such as the salad bar.

22. What are the cafeteria hours?

_____

_____

_____

_____

_____

23. How many customers are served daily?

_____

_____

What percentage of hospital employees use the cafeteria?

_____

_____

What is the average check?

_____

_____

24. Are special promotions or themed events planned for the cafeteria? If so, give two examples.

_____

_____

_____

_____

25. What are the most popular menu items?

_____

_____

_____

_____

_____

26. How are the foods garnished and kept looking fresh?

_____

_____

_____

_____

27. How much time do healthcare employees have to eat lunch or dinner?

_____

28. Is there much local competition from restaurants and fast food establishments?

_____

_____

_____

_____

29. How are cafeteria prices set? Are prices subsidized?

_____

_____

_____

Is there one pricing structure for all customers? If not, describe the pricing structure.

_____

30. Describe other nonpatient foodservices offered, such as in-house catering, take-home foods, vending, or outside catering.

_____

_____

_____

_____

_____

_____

_____

Nutrition Services

31. Describe the types of nutrition services offered.

_____

_____

_____

_____

_____

_____

32. What are the job titles of employees providing nutrition services?

_____

_____

_____

_____

33. For each job title, what are the three most important functions?

_____

_____

_____

_____

_____

_____

_____

_____

_____

_____

34. Sketch below the organizational chart of those providing nutrition services.

35. Find out from a registered dietitian (R.D.) on staff the qualifications one must meet to become an R.D.

_____

_____

_____

_____

_____

_____

36. Ask the director or assistant director for his or her thoughts on three current concerns and issues facing health care foodservice.

_____

_____

_____

_____

_____

_____

_____

_____

_____

37.

38.

Name: _____

Date(s): _____

Department/Location: _____

Source of Information: _____

1. How many students attend the college?

   _____

   _____

   How many are residents?

   _____

   _____

   How many are commuters?

   _____

   _____

2. Describe common characteristics of the students, such as their ages, mostly male or female, and so on.

   _____

   _____

   _____

   _____

3. Describe the meal plan options available to both resident and commuter students. Do resident students have to buy a meal plan?

   _____

   _____

   _____

   _____

   _____

   _____

   _____

   _____

4. What dining services are offered, such as cafeterias, restaurant, catering, vending, and so on?

_____

_____

_____

_____

Who uses each of these services?

_____

_____

_____

5. Which dining service is the most heavily used?

_____

_____

6. What is the average daily number of transactions for each dining service area?

_____

_____

_____

_____

7. What is the participation rate?

_____

_____

8. What is the average check for each dining service area?

_____

_____

_____

_____

9. Which menu items are most popular?

_____

_____

_____

_____

_____

How long is the menu cycle? **Attach** *a menu if possible.*

_____

10. Does the foodservice use many brand-name foods and beverages? Describe them.

_____

_____

_____

_____

_____

Are the brands used national brands and/or in-house brands?

_____

_____

_____

_____

11. Arc special promotions or themed events planned for the cafeteria? If so, give two examples.

_____

_____

_____

_____

_____

12. What special student needs do you meet, such as vegetarian entrées or training tables for athletes?

_____

_____

_____

_____

13. How are prices for the meals and meal plans determined?

_____

_____

_____

_____

14. What is an average cost per student meal?

Food cost _____

Labor cost _____

15. Who works in the college and university foodservice besides students?

_____

_____

_____

_____

16. Ask the director or assistant director for his or her thoughts on three current concerns and issues facing college and university foodservice.

_____

_____

_____

_____

_____

_____

_____

_____

_____

_____

17.

18.

Name: _____

Date(s): _____

Department/Location: _____

Source of Information: _____

1. How many residents live in independent-living housing?

_____

_____

   How many independent-living residences have been built?

_____

_____

2. Is assisted-living housing available? If so, how many beds?

_____

   Are skilled-nursing beds available? If so, how many beds?

_____

3. Do residents own or rent their housing?

_____

_____

   How much does it cost to own or rent here?

_____

_____

_____

_____

_____

4. How many meals per day or per week are residents entitled to?

_____

_____

_____

_____

5. What characteristics, such as age, do many residents have in common?

_____

_____

_____

_____

6. What services, such as 24-hour nursing, are offered to the residents?

_____

_____

_____

_____

_____

7. What dining services are offered, such as restaurant, catering, and vending?

_____

_____

_____

_____

_____

8. Which dining service is the most heavily used?

_____

_____

9. What is the average daily number of meals served?

_____

_____

_____

_____

10. Which menu items are most popular in the main dining room?

_____

_____

_____

_____

_____

How often does the menu change? Explain. **Attach** *a menu if possible.*

_____

_____

_____

_____

11. Are special promotions or themed events planned? If so, give two examples.

_____

_____

_____

_____

12. What do the average food cost and average labor cost run?

_____

_____

13. Ask the director or assistant director for his or her thoughts on three current concerns and issues facing retirement community foodservice.

_____

_____

_____

_____

_____

_____

_____

_____

14.

15.

5

Training Goals and Training Guides for Hospitality Human Resource Management

This section includes Training Goals and Training Guides for the following functions in hospitality human resources management:

- Staffing
- Training
- Discipline
- Performance evaluation

Depending on the size and organization of your company, one or more people might work full-time in human resources. In many smaller companies without a human resource department, managers assume these functions.

Whatever the case, read the Training Goals carefully and fill in the Training Guides as accurately and completely as possible. Don't feel embarrassed to ask questions of employees and supervisors. You'll soon find out that most employees are delighted to tell you about their jobs.

Before starting to use the Training Goals and Training Guides, check with your instructor to see if you should complete the Introductory Worksheet that starts this chapter: Introduction to Human Resource Management. This worksheet is designed to be used initially to help you see the "big picture." Good luck!

Name: _____

Date(s): _____

Department/Location: _____

Source of Information: _____

1. Sketch below the organizational chart for the human resource area, or **attach** a copy of the chart.

2. List the major duties of each job in human resources.

_____

_____

_____

_____

_____

_____

_____

_____

_____

_____

_____

_____

_____

_____

3. How many employees are in the company?

_____

_____

4. What is the average turnover rate?

_____

_____

How is it calculated?

_____

_____

_____

_____

5. Does a union (or unions) represent any of the company's employees? If so, identify the union and the group of employees it represents.

_____

_____

_____

_____

6.

7.

1. Identify staffing needs.

2. Use job descriptions and job specifications in staffing.

3. Identify appropriate internal and external recruiting methods and sources for open positions.

4. Write a classified "Help Wanted" advertisement.

5. Evaluate completed application forms.

6. Conduct an interview.

7. Administer preemployment tests.

8. Perform a reference check.

9. Make selection decisions.

10. Make job offers.

11. Explain benefits available to full-time and part-time employees.

12. Explain what taxes and other amount are withheld from employees' paychecks.

13. _____

14. _____

Name: _____

Date(s): _____

Department/Location: _____

Source of Information: _____

Recruiting and Selecting

1. Does your company forecast the number of workers needed for various job titles? If so, describe the process.

   _____

   _____

   _____

   _____

   _____

   _____

   _____

   _____

2. What are the principal sources and methods for recruiting prospective employees?

   _____

   _____

   _____

   _____

   _____

   _____

   _____

   _____

   _____

   _____

   _____

   _____

   _____

Does the company use classified advertising in area newspapers? If so, write down a typical ad.

_____

_____

_____

_____

_____

_____

_____

Which sources and methods seem to produce the best results?

_____

_____

_____

_____

_____

3. Are employment agencies used at all to find employees?

_____

_____

Why or why not?

_____

_____

_____

_____

If agencies are used, how is the fee handled?

_____

_____

_____

For which jobs is the employment agency used?

_____

_____

_____

4. Are current employees informed of job openings within the company and encouraged to apply if they meet the qualifications? Why or why not?

_____

_____

_____

_____

_____

_____

_____

5. Are there written job descriptions for each job that are used to recruit and select employees?

_____

_____

6. Are there job specifications for each job that are used to recruit and select employees?

_____

_____

7. Do applicants have to fill out an employment application? If so, **attach** a copy of the application.

_____

8. Who interviews applicants? Include titles.

_____

_____

_____

9. Describe the general interview process used here.

_____

_____

_____

_____

_____

_____

Where are the interviews conducted?

_____

_____

Who takes part in the interviews?

_____

_____

10. Do potential applicants for any jobs have to undergo testing? If so, describe the testing.

_____

_____

_____

_____

_____

_____

11. What type of reference check is done on a prospective employee? If a form is used, **attach** it. If a form is not used, what are the typical questions asked in a telephone interview?

_____

_____

_____

_____

_____

_____

_____

12. Once an applicant has been selected, how is the job offer extended?

_____

_____

_____

13. Once employees are selected and hired, do they have to undergo a physical examination? Why or why not?

_____

_____

_____

_____

14. Ask someone in human resources or management what he or she looks for when selecting applicants.

_____

_____

_____

_____

15. Ask someone in human resources or management about the legal issues that need to be taken into account in staffing.

_____

_____

_____

_____

_____

_____

_____

_____

_____

_____

Compensation

16. What are the current federal regulations concerning wages and hours? Include minimum wage and overtime information.

_____

_____

_____

_____

_____

_____

_____

17. What are the current federal regulations concerning tips?

_____

_____

_____

_____

_____

_____

_____

18. List the taxes and other amounts withheld from employee paychecks. Also list the rate at which each is withheld.

_____

_____

_____

_____

_____

_____

_____

19. Describe the benefits package for full-time employees. Include paid time off, health insurance, and so forth.

_____

_____

_____

_____

_____

_____

_____

_____

_____

_____

20. What kinds of benefits do part-timers receive?

_____

_____

_____

_____

_____

_____

_____

_____

21.

22.

1. Conduct new employee orientation.

2. Decide on appropriate training methods for job training and retraining.

3. Train a new employee on how to do his or her job.

4. _____

5. _____

Name: _____

Date(s): _____

Department/Location: _____

Source of Information: _____

1. How are new employees oriented to their new jobs and the company? If there is a formal orientation, list its major components. **Attach** an orientation checklist if available.

_____

_____

_____

_____

_____

_____

_____

_____

_____

_____

Who conducts new employee orientation?

_____

_____

2. List at least six rules of conduct, dress, or behavior that are discussed during orientation.

_____

_____

_____

_____

_____

_____

_____

_____

_____

_____

3. Are employees given a copy of an employee handbook at this time? Is one available?

_____

_____

4. After an employee has been oriented, describe the typical job training that takes place.

_____

_____

_____

_____

_____

_____

5. Give examples in your operation of when group training is used and when individual training is used.

_____

_____

_____

_____

_____

_____

_____

_____

_____

6. Which of the following training methods or approaches are used in your operation?

On-the-job training _____

Lecture and discussion _____

Demonstration _____

Videotapes _____

Roleplaying _____

Games _____

Other: _____

7. Is there an ongoing training program in your operation, or is training limited to new employees?

_____

_____

_____

8. When an employee's performance drops below par, who is responsible for retraining the employee?

_____

_____

_____

_____

_____

9. Ask a supervisor to enumerate the advantages and disadvantages of training. Which outweighs the other?

_____

_____

_____

_____

_____

_____

_____

_____

_____

10.

11.

1. Explain how to properly use the company's discipline policy and procedure.

2. Maintain discipline records.

3. Conduct a disciplinary interview.

4. Describe when termination is appropriate.

5. _____

6. _____

Name: _____

Date(s): _____

Department/Location: _____

Source of Information: _____

1. Is there a discipline policy and procedure available? If yes, **attach** a copy or describe its major parts.

_____

_____

_____

_____

_____

_____

_____

_____

_____

_____

_____

_____

Are there elements of negative discipline, positive discipline, and/or progressive discipline in the policy?

_____

_____

_____

_____

_____

_____

2. Is any type of warning system used with employees who are not doing their job properly?

_____

_____

_____

If so, what is the system? How many warnings are there? Is each warning verbal or written?

_____

_____

_____

_____

_____

_____

Is the system used as it is designed?

_____

_____

_____

_____

3. Are any infractions cause for immediate dismissal? If so, list them.

_____

_____

_____

_____

_____

_____

_____

4. Is excessive absenteeism or tardiness defined? If so, give the definition.

_____

_____

_____

_____

_____

_____

5. What group of employees accounts for the highest absenteeism? (Think of age, sex, or job classification.)

_____

_____

_____

What are some common causes of absenteeism?

_____

_____

_____

_____

Do certain employees have a habit of calling in absent on the day before or after a day or weekend off?

_____

_____

6. How are attendance records maintained? **Attach** any pertinent forms.

_____

_____

_____

_____

_____

7. When a disciplinary incident, such as an employee smoking on the job, occurs, what are the typical steps a supervisor or manager will go through?

_____

_____

_____

_____

_____

_____

_____

_____

_____

_____

8. What happens to an employee who is doing the job poorly and is suspected to be working under the influence of alcohol?

_____

_____

_____

_____

_____

_____

_____

_____

9. Are employees ever transferred to another job because they are not doing well in the job which they are assigned?

_____

_____

If yes, how has this worked out?

_____

_____

_____

_____

If no, what is done with the person?

_____

_____

_____

_____

10. Under what conditions can an employee be terminated?

_____

_____

_____

_____

_____

_____

_____

_____

11. What is the cause most often given as the reason for the discharge of an employee?

_____

_____

_____

_____

_____

12. Are employees praised, both verbally and in writing, for positive actions they take? Describe the praise.

_____

_____

_____

_____

_____

13. Ask a supervisor to identify four essential elements of successful discipline and explain the importance of each.

_____

_____

_____

_____

_____

_____

_____

_____

_____

_____

_____

_____

14.

15.

1. Complete performance evaluation forms fairly and objectively.

2. Conduct a performance evaluation interview.

3. Discuss how to evaluate employees between performance evaluation times.

4. _____

5. _____

Name: _____

Date(s): _____

Department/Location: _____

Source of Information: _____

1. Is any type of rating or evaluation done on the employee on a scheduled basis?

   _____

   _____

2. Is the performance evaluation used as the basis for raises or promotions?

   _____

   _____

   _____

3. Describe the forms used in evaluating employees by answering the following questions. **Attach** a form if available.

   On which performance dimensions or categories is the employee rated?

   _____

   _____

   _____

   _____

   _____

   _____

   _____

   Explain the rating scale used.

   _____

   _____

   _____

   _____

   _____

   _____

Is the form easy to use? Why or why not?

_____

_____

_____

_____

Is the form fair and objective? Why or why not?

_____

_____

_____

_____

_____

_____

4. Explain how a supervisor or manager completes the form and discusses it with the employee.

_____

_____

_____

_____

_____

_____

_____

_____

_____

_____

_____

5. Are employees asked to do a self-appraisal as part of the performance evaluation process? If so, describe.

_____

_____

_____

_____

_____

6. Ask someone who routinely performs performance evaluations to discuss possible pitfalls in this process, from filling in the form to completing the performance evaluation discussion with the employee.

_____
_____
_____
_____
_____
_____
_____
_____
_____
_____

7. Once a performance evaluation has been completed, where does the form go?

_____
_____
_____
_____
_____
_____

Does the employee get a copy?

_____
_____
_____

8. How is an employee evaluated between performance evaluations?

_____
_____
_____
_____
_____
_____
_____
_____

9. What are the opportunities for career advancement for entry-level employees? Discuss the opportunities for at least two different entry-level positions within your company.

_____

_____

_____

_____

_____

_____

_____

_____

_____

_____

10.

11.